A Panorama of Statistics

A Panorama of Statistics

Perspectives, puzzles and paradoxes in statistics

Eric Sowey

The University of NSW, Sydney, Australia

Peter Petocz

Macquarie University, Sydney, Australia

This edition first published 2017
© 2017 by John Wiley & Sons, Ltd

Registered Office
John Wiley & Sons, Ltd, The Atrium, Southern Gate, Chichester, West Sussex, PO19 8SQ,
United Kingdom

Editorial Offices
9600 Garsington Road, Oxford, OX4 2DQ, United Kingdom
The Atrium, Southern Gate, Chichester, West Sussex, PO19 8SQ, United Kingdom

For details of our global editorial offices, for customer services and for information about
how to apply for permission to reuse the copyright material in this book please see our
website at www.wiley.com/wiley-blackwell.

The right of each of the two authors to be identified as an author of this work has been
asserted in accordance with the UK Copyright, Designs and Patents Act 1988.

Library of Congress Cataloging-in-Publication data applied for

ISBN: 9781119075820

A catalogue record for this book is available from the British Library.

Wiley also publishes its books in a variety of electronic formats. Some content that appears in
print may not be available in electronic books.

Cover image: Jeffrey Smart, Container train in landscape (1983–84), Arts Centre Melbourne,
Australia. Reproduced by permission of the Estate of Jeffrey Smart.

Set in 10/12.5pt Warnock by SPi Global, Pondicherry, India

Printed and bound in Malaysia by Vivar Printing Sdn Bhd

10 9 8 7 6 5 4 3 2 1

ES   *To Anne, Helen, Michelle and David*

PP   *To Anna*

Contents

Preface

Most people would cheerfully agree that statistics is a useful subject – but how many would recognise that it also has many facets which are engaging, and even fascinating?

We have written this book for students and their teachers, as well as for practitioners – indeed, for anyone who knows some statistics. If this describes you, we invite you to come with us on a panoramic tour of the subject. Our intent is to highlight a variety of engaging and quirky facets of statistics, and to let you discover their fascinations. Even if you are only casually acquainted with statistical ideas, there is still much in this book for you.

This is not a textbook. In a lively way, it expands understanding of topics that are outside the scope of most textbooks – topics you are unlikely to find brought together all in the one place elsewhere.

Each of the first 25 chapters is devoted to a different statistical theme. These chapters have a common structure. First, there is an Overview, offering perspectives on the theme – often from several points of view. About half of these Overviews need, as quantitative background, only high school mathematics with a basic foundation in statistics. For the rest, it may be helpful to have completed an introductory college or university course in statistics.

Following the Overview, each chapter poses five questions to pique your curiosity and stimulate you to make your own discoveries. These questions all relate to the theme of the chapter. As you seek answers to these questions, we expect you will be surprised by the variety of ways in which statistics can capture and hold your interest.

The questions are not for technical, numerical or web-search drill. Rather, they seek to widen your knowledge and deepen your insight. There are questions about statistical ideas and probabilistic thinking, about the value of statistical techniques, about both innocent and cunning misuses of statistics, about pathbreaking inspirations of statistical pioneers, and about making the best practical use of statistical information. Also, there are amusing statistical puzzles to unravel and tantalising statistical paradoxes to resolve. Some questions have a single correct answer, but many more invite your reflection and your exploration of alternatives.

We invite you to plunge in and tackle the questions that you find appealing. Compare your findings with our own answers (including wide-ranging commentary), which are collected together in CHAPTER 26.

To help you to choose questions that best match your current statistical background, we have labelled each question A, B or C. Questions labelled A (40% of the total) are well suited to those who are studying an introductory course in statistics at tertiary level. Good senior high school students should also find many of them within their capability. Questions labelled B (55% of the total) cover a wide spectrum of challenges and, in many cases, a knowledge of statistics at the level of a second course will make the best starting point. The remaining 5% of questions are labelled C, and are for graduates in statistics, including professional practitioners.

In each chapter, the Overview and its five questions are extensively cross-referenced to related material in other chapters. They also include suggestions for further reading, both in print and online. The web links are available live on this book's companion website www.wiley.com/go/sowey/ apanoramaofstatistics.

To make it clear when a mentioned Chapter, Question or Figure refers to a place elsewhere in this book (rather than to an external source), the words are printed in small capitals: CHAPTER, QUESTION, FIGURE.

We hope that your time spent with this book will be enjoyable and enriching.

May 2016

Eric Sowey
Peter Petocz

Acknowledgments

Our longest-standing debt is to Martin Gardner (1914–2010), whose monthly 'Mathematical Games' columns appeared in *Scientific American* from 1956 to 1981. We were still in high school when we first discovered these short, lively and often quirky essays, which have drawn tens of thousands of people (us included) in to the pleasures and fascinations of mathematics.

Among the approximately 300 essays (which were collected into 15 books between 1959 and 1997), there are fewer than a dozen on probability, and just a handful in which a statistical concept is mentioned. Each of us had, in the past, wondered why no-one had yet brought Gardner's vibrant approach to statistics. When we discovered our common interest, it became an irresistible challenge to do it ourselves. Our thirty-six columns of 'Statistical Diversions' – conceived very much in the spirit of Martin Gardner – appeared between 2003 and 2015 in each issue of *Teaching Statistics*, published by Wiley, both online (at http://onlinelibrary.wiley.com/journal/10.1111/(ISSN)1467-9639) and in hard copy.

This book represents a substantial revision, reorganisation and extension of the material in those columns. We celebrate the memory of Martin Gardner by citing four of his marvellous essays in our answers to QUESTIONS 10.5, 11.4 and 12.2 in this book.

We warmly thank the Editors of *Teaching Statistics* from 2003 to 2015 – Gerald Goodall, Roger Johnson, Paul Hewson and Helen MacGillivray – for their editorial encouragement and advice as they welcomed each of our Statistical Diversions columns.

John Aldrich, Geoff Cumming, Gerald Goodall and Jan Kmenta each read several chapters of this book. Tim Keighley and Agnes Petocz read the book through in its entirety. All these colleagues willingly responded to our invitation, bringing to the task their deep knowledge of statistics and its application in other disciplines and their cultural vantage points in Australia, the UK and the USA. We are grateful to them for the multiple ways in which their comments have improved this book.

The cover reproduces a striking panoramic painting titled *Container train in landscape* (1983–84, oil on five hardboard panels, 113.5 × 985 cm). This work, almost ten metres long, hangs in the Arts Centre Melbourne and is the gift of Mark and Eva Besen. It is by the Australian artist Jeffrey Smart (1921–2013), who lived for many years in Tuscany and whose paintings have captured worldwide attention. Smart's works are admired for, among other features, their vivid and arresting colours and their geometrically precise composition. We are delighted that the Estate of Jeffrey Smart has given permission for the use of the cover image.

David Urbinder's original drawings cheerily embellish our pages. We reproduce them by permission. Michelle Sowey's creativity guided us on the cover design.

Neville Davies, Ruma Falk, Rob Gould, Bart Holland, Larry Lesser, Milo Schield, Nel Verhoeven and Graham Wood made valued contributions while we were creating our Statistical Diversions columns, or in the course of preparing this book for publication.

We thank the following people for permission to reproduce material in the Figures: Michael Rip (FIGURE 1.2), Gavan Tredoux (FIGURE 18.1) and David Thomas (FIGURE 21.1).

To our Editors at Wiley, Debbie Jupe, Heather Kay, Shivana Raj, Alison Oliver, Uma Chelladurai and Brian Asbury we express our appreciation for their advice, guidance and support throughout the publication process.

Part I

Introduction

1

Why is statistics such a fascinating subject?

In the real world, little is certain. Almost everything that happens is influenced, to a greater or lesser degree, by chance. As we shall explain in this chapter, statistics is our best guide for understanding the behaviour of chance events that are, in some way, measurable. No other field of knowledge is as vital for the purpose. This is quite a remarkable truth and, statisticians will agree, one source of the subject's fascination.

You may know the saying: data are not information and information is not knowledge. This is a useful reminder! Even more useful is the insight that it is statistical methods that play the major role in turning data into information and information into knowledge.

In a world of heavily promoted commercial and political claims, a familiarity with statistical thinking can bring enormous personal and social benefits. It can help everyone to judge better what claims are trustworthy, and so become more competent and wiser as citizens, as consumers and as voters. In short, it can make ours not only a more numerate, but also a more accurately informed, society. This is an ideal we shall return to in CHAPTER 3.

Chance events are studied in the physical, biological and social sciences, in architecture and engineering, in medicine and law, in finance and marketing, and in history and politics. In all these fields and more, statistics has well-established credentials. To use John Tukey's charming expression, 'being a statistician [means] you get to play in everyone's backyard'. (There is more about this brilliant US statistician in CHAPTER 22, FIGURE 22.2.)

---oOo---

To gain a bird's eye view of the kinds of practical conclusions this subject can deliver, put yourself now in a situation that is typical for an applied statistician.

A Panorama of Statistics: Perspectives, Puzzles and Paradoxes in Statistics, First Edition.
Eric Sowey and Peter Petocz.
© 2017 John Wiley & Sons, Ltd. Published 2017 by John Wiley & Sons, Ltd.
Companion website: www.wiley.com/go/sowey/apanoramaofstatistics

Suppose you have collected some data over a continuous period of 150 weekdays on the daily number of employees absent from work in a large insurance company. These 150 numbers will, at first, seem to be just a jumble of figures. However, you – the statistician – are always looking for patterns in data, because patterns suggest the presence of some sort of systematic behaviour that may turn out to be interesting. So you ask yourself: can I find any evidence of persisting patterns in this mass of figures? You might pause to reflect on what sorts of meaningful patterns might be present, and how you could arrange the data to reveal each of them. It is clear that, even at this early stage of data analysis, there is lots of scope for creative thinking.

Exercising creativity is the antithesis of following formalised procedures. Unfortunately, there are still textbooks that present statistical analysis as no more than a set of formalised procedures. In practice, it is quite the contrary. Experience teaches the perceptive statistician that a sharpened curiosity, together with some preliminary 'prodding' of the data, can often lead to surprising and important discoveries. Tukey vigorously advocated this approach. He called it 'exploratory data analysis'. Chatfield (2002) excellently conveys its flavour.

In this exploratory spirit, let's say you decide to find out whether there is any pattern of absenteeism across the week. Suppose you notice at once that there seem generally to be more absentees on Mondays and Fridays than on the other days of the week. To confirm this impression, you average the absentee numbers for each of the days of the week over the 30 weeks of data. And, indeed, the averages are higher for Mondays and Fridays.

Then, to sharpen the picture further, you put the Monday and Friday averages into one group (Group A), and the Tuesday, Wednesday and Thursday averages into a second group (Group B), then combine the values in each group by averaging them. You find the Group A average is 104 (representing 9.5% of staff) and the Group B average is 85 (representing 7.8% of staff).

This summarisation of 30 weeks of company experience has demonstrated that staff absenteeism is, on average, 1.7 percentage points higher on Mondays and Fridays as compared with Tuesdays, Wednesdays and Thursdays. Quantifying this difference is a first step towards better understanding employee absenteeism in that company *over the longer term* – whether your primary interest is possible employee discontent, or the financial costs of absenteeism to management.

Creating different kinds of data summaries is termed *statistical description.* Numerical and graphical methods for summarising data are valuable, because they make data analysis more manageable and because they can reveal otherwise unnoticed patterns.

Even more valuable are the methods of statistics that enable statisticians to generalise to a wider setting whatever interesting behaviour they may have detected in the original data. The process of generalisation in the face of the uncertainties of the real world is called *statistical inference.* What makes a statistical generalisation so valuable is that it comes with an objective measure of the likelihood that it is correct.

Clearly, a generalisation will be useful in practice only if it has a high chance of being correct. However, it is equally clear that we can never be sure that a generalisation is correct, because uncertainty is so pervasive in the real world.

To return to the example we are pursuing, you may be concerned that the pattern of absenteeism detected in 30 weeks of data might continue indefinitely, to the detriment of the company. At the same time, you may be unsure that that pattern actually is a long-term phenomenon. After all, it may have appeared in the collected data only by chance. You might, therefore, have good reason to widen your focus, from absenteeism in a particular 30-week period to absenteeism in the long term.

You can test the hypothesis that the pattern you have detected in your data occurred by chance alone against the alternative hypothesis that it did *not* occur by chance alone. The alternative hypothesis suggests that the pattern is actually persistent – that is, that it is built into the long-term behaviour of the company if there are no internal changes (by management) or external impacts (from business conditions generally). As just mentioned, the statistical technique for performing such a hypothesis test can also supply a measure of the likelihood that the test result is correct. For more on hypothesis testing, see CHAPTER 16.

When you do the test, suppose your finding is in favour of the alternative hypothesis. (Estimating the likelihood that this finding is correct requires information beyond our scope here, but there are ways of testing which optimise that likelihood.) Your finding suggests a long-term persisting pattern in absenteeism. You then have grounds for recommending a suitable intervention to management.

Generalising to 'a wider setting' can also include to 'a future setting', as this example illustrates. In other words, statistical inference, appropriately applied, can offer a cautious way of forecasting the future – a dream that has fascinated humankind from time immemorial.

In short, statistical inference is a logical process that deals with 'chancy' data and generalises what those data reveal to wider settings. In those wider settings, it provides precise (as opposed to vague) conclusions which have a high chance of being correct.

---oOo---

But this seems paradoxical! What sort of logic is it that allows highly reliable conclusions to be drawn in the face of the world's uncertainties? (Here, and in what follows, we say 'highly reliable' as a shorter way of saying 'having a high chance of being correct'.)

To answer this pivotal question, we need first to offer you a short overview of the alternative *systems of logic* that philosophers have devised over the centuries. For an extended exposition, see Barker (2003).

A system of logic is a set of rules for reasoning from given assumptions towards reliable conclusions. There are just two systems of logic: *deduction* and *induction*. Each system contains two kinds of rules:

i) rules for drawing precise conclusions in all contexts where that logic is applicable; and
ii) rules for objectively assessing how likely it is that such precise conclusions are actually correct.

The conclusions that each system yields are called deductive inferences and inductive inferences, respectively.

It's worth a moment's digression to mention that there are two other thought processes – analogy and intuition – which are sometimes used in an attempt to draw reliable conclusions. However, these are not systems of logic, because they lack rules, either of the second kind (analogy) or of both kinds (intuition). Thus, conclusions reached by analogy or by intuition are, in general, less reliable than those obtained by deduction or induction. You will find in QUESTIONS 9.4 and 9.5, respectively, examples of the failure of analogy and of intuition.

In what kind of problem setting is deduction applicable? And in what kind of setting is induction applicable? The distinguishing criterion is whether the setting is (or is assumed to be) one of complete certainty.

In a setting of complete certainty, deduction is applicable, and there is no need for induction. Why? Because if all assumptions made (including the assumption that nothing is uncertain) are correct, and the rules of deduction are obeyed, then a deductive inference *must* be correct.

If you think back to the problems you solved in school mathematics (algebra, calculus, geometry and trigonometry), you will recall that, in these areas, chance influences were given no role whatever. No surprise, then, that deduction is the system of logic that underpins all mathematical inferences – which mathematicians call 'theorems'.

It is a great strength of deductively based theorems that they are *universally* correct (i.e. for every case where the same assumptions apply). For instance, given the assumptions of (Euclidean) plane geometry and the definition of a right-angled triangle, Pythagoras's Theorem is true for every such triangle, without exception.

Now, what about reasoning in a setting of uncertainty? Here, induction is applicable, and you can see how it contrasts with deduction. In a setting of uncertainty, even if all assumptions made are correct and the rules of induction are obeyed, an inductive inference *might not* be correct, because of chance influences, which are always at work.

Still, induction is more reliable in this setting than deduction, because the rules of induction explicitly recognise the influence of chance, whereas the rules of deduction make no mention of it whatever. In short, when the influence of chance is inescapable – as is the case in most real-world situations – induction is the system of logic that underpins all inferences.

If you head out one morning at the usual time to catch your regular 7.30 am train to work, you are reasoning inductively (or 'making an inductive inference'). Train timetables are vulnerable to bad weather delays, signal failures, and accidents along the rail line. So, even if, on all previous occasions, the 7.30 am train arrived on time, it is *not* correct to conclude that it *must* arrive on time today. Of course, the train is *highly likely* to arrive on time. But you cannot logically say more than that.

It follows that inductive inferences that are highly reliable in one circumstance are not necessarily highly reliable in other circumstances, *even where the same assumptions apply*. That is because chance influences can take many different forms, and always (by definition) come 'out of the blue'. For instance, even though the on-time arrival of your 7.30 am train has turned out to be highly reliable, reliability may shrink when you are waiting for your train home in the afternoon peak hours – the most likely period (our Sydney experience shows) in which unforeseen disruptions to train schedules occur.

---oOo---

We have now seen that it is inductive logic that enables inferences to be made in the face of uncertainty, and that such inferences need not be reliable in any particular instance. You may be thinking, 'it's no great achievement to produce unreliable conclusions'.

This thought prompts a new question: given that induction is the only system of logic that is applicable in chance situations, can rules of induction be configured to allow the conclusions it produces to be highly reliable in principle?

The answer is yes. Over the past century, statisticians have given a great deal of attention to refining the rules of induction that have come down to us through the cumulative work of earlier logicians, beginning with Francis Bacon (1561–1626). These refined rules of induction, now designed expressly for quantitative inferences, are called the rules of statistical induction.

The distinction between an inductive inference and a statistical inductive inference may seem both subtle and trivial. For an excellent discussion of the distinction, see chapters 4 and 5 of Burbidge (1990). While it is a subtle distinction, it is definitely not trivial. Relative to alternative ways of specifying rules of induction, the rules of statistical induction have, in principle, the highest chance of producing reliable conclusions in any particular instance.

In other words, *statistical inductive inference is the most reliable version of the most powerful logic that we have for reasoning about chance events.* As statisticians, we find this both fascinating and inspiring.

For simplicity, we shall now drop the formal term 'statistical inductive inference' and revert to using its conventional equivalent – 'statistical inference'.

Statistical description and statistical inference are the workaday roles of statistics. These two roles define the highways, so to speak, of statistical activity.

---oOo---

Statistics also has many byways. You will find them prominent in this book. Yet, they are not front and centre in statistics education curricula, nor are they part of the routine activities of applied statisticians. So, how do they come to attention?

Statistical theorists come upon several of these byways when refining and enhancing basic methods of analysis. In one category are *paradoxes of probability and statistics* (see, for some examples, CHAPTERS 10 and 11). In another are *problems of using standard statistical techniques in non-standard situations* (see CHAPTER 17). In a third are *unifying principles*: fundamental ideas that are common to diverse areas of statistical theory. Discovering unifying principles means identifying previously unrecognised similarities in the subject. Unification makes the subject more coherent, and easier to understand as a whole. Examples of unifying principles are the Central Limit Theorem (see CHAPTERS 12 and 14) and the power law (see CHAPTER 24).

Another byway is *the history of statistical ideas.* Historians with this special interest bring to life the philosophical standpoints, the intellectual explorations and the (sometimes controversial) writings of statistical pioneers, going back over several centuries. Though pioneering achievements may look straightforward to us in hindsight, the pioneers generally had to struggle to succeed – first, in formulating exactly what it was they were trying to solve, and then in harnessing all their insight, knowledge and creativity towards finding solutions, often in the face of sceptical critique (see, for instance, CHAPTERS 18 and 22).

Yet another byway is *the social impact of statistics.* Here are three paths worth exploring on this byway: consequences of the low level of statistical literacy in the general community, and efforts to raise it (see CHAPTERS 3 and 6); public recognition of statisticians' achievements via eponymy (see CHAPTER 23); and the negative effects of widespread public misuse of statistical methods, whether from inexperienced analysts' ignorance, or from a deliberate intention to deceive (see CHAPTERS 8 and 9).

These are by no means all the byways of statistics. You will discover others for yourself, we hope, scattered through the following chapters.

Highways and byways.

You may then also come to share our view that, to people who are curious, the lightly visited byways of statistics can be even more delightful, more surprising and more fascinating than the heavily travelled highways of standard statistical practice.

If, at this point, you would like to refresh your knowledge of statistical ideas and principles, we recommend browsing the following technically very accessible books: Freedman, Pisani and Purves (2007), and Moore and Notz (2012).

Questions

Question 1.1 (A)

FIGURE 1.1 shows, on a logarithmic horizontal scale, the cumulative percentage frequency of heads in a sequence of 10,000 tosses of a coin.

These 10,000 tosses were performed by a South African statistician, John Kerrich, who went on to be the Foundation Professor of Statistics at Witwatersrand University in 1957.

a) Where, and under what unusual circumstances, did Kerrich perform these 10,000 tosses?
b) Does the information in the graph help us to define 'the probability of getting a head when a fair coin is tossed once'?

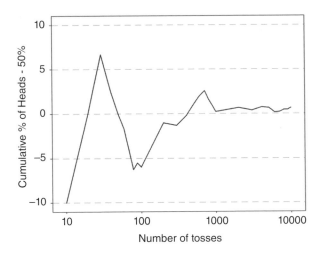

Figure 1.1 Scatterplot of Kerrich's coin-tossing results. Data from Freedman, Pisani and Purves (2007).

Question 1.2 (A)

When young children are asked about their understanding of probability, they quickly decide that the sample space for rolling a single die consists of six equally likely outcomes. When it comes to two dice, however, they often conclude that the sample space has 21 outcomes that are equally likely. Where does the number 21 come from?

Question 1.3 (A)

'Most people in London have more than the average number of legs.' Is this statement correct? Does it indicate some misuse of statistical methods?

Question 1.4 (A)

Based on thirty continuous years of recorded temperature data, the average temperature over the 12 months in a calendar year in New York is 11.7 °C, in New Delhi it is 25.2 °C, and in Singapore it is 27.1 °C. (To see the data – which may vary slightly over time – go online to [1.1], select the three cities in turn from the menu, and find the monthly average temperatures in the left-hand frame for each city.)

Does this mean that it gets roughly twice as hot in New Delhi during the year as it does in New York? Does it mean that the climate in Singapore is much the same as that in New Delhi?

Question 1.5 (B)

The map in FIGURE 1.2 shows a part of London. By whom was it drawn and when? With what famous event in the history of epidemiology is it connected? (*Hint*: note the street-corner pump.)

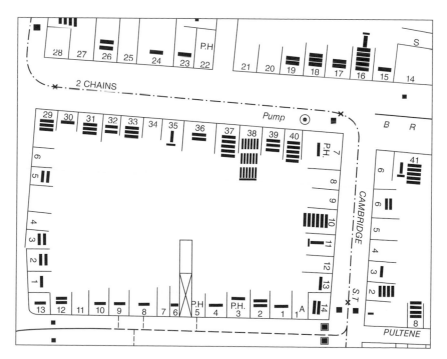

Figure 1.2 Extract from a map showing a part of London. Reproduced with the permission of Michael Rip.

References

Print

Barker, S.F. (2003). *The Elements of Logic*, 6th edition. McGraw-Hill.
Burbidge, J. (1990). *Within Reason: A Guide to Non-Deductive Reasoning.* Broadview Press.
Chatfield, C. (2002). Confessions of a pragmatic statistician. *The Statistician* **51**, 1–20.
Freedman, D., Pisani, R. and Purves, R. (2007). *Statistics*, 4th edition. Norton.
Moore, D.S. and Notz, W.I. (2012). *Statistics: Concepts and Controversies*, 8th edition. Freeman.

Online

[1.1] www.eurometeo.com/english/climate

2

How statistics differs from mathematics

'What's the difference between mathematics and statistics?' Many school students put this question to their teacher, aware that these subjects are related but not clear on what it is, exactly, that distinguishes them. Unravelling this puzzle is generally not made any easier for students by the fact that, in most schools around the world, it is the mathematics department that normally teaches statistics. To these curious but bewildered students, 'maths' seems to be defined by the topics that the teacher and the textbook say are maths, and similarly for 'stats'. So, algebra, calculus, geometry and trigonometry are 'maths', while frequency distributions, averages, sampling, the normal distribution, and estimation are 'stats'. That doesn't go very far towards providing a convincing answer to our opening question. Anyway, what about probability? Is that 'maths' or 'stats'?

A thoughtful teacher will want to supply a better answer. Surprisingly, in our experience, a better answer is rarely found either in curriculum documents or in textbooks. So let's see if we can formulate a better answer in a straightforward way.

A constructive start is to ask in what ways statistics problems differ from mathematics problems.

Here is something fairly obvious: statistics problems have a lot to do with getting a view of the variability in data collected from the real world. For example, a statistical problem may present 100 measurements (by different people) of the length of a particular object, using a tape measure, with the assigned task being to construct a frequency distribution of these measurements to see whether measurement errors tend to be symmetrical about the correct value, or whether people tend to veer more to one side or the other. By contrast, in a mathematical problem involving the length of an object,

A Panorama of Statistics: Perspectives, Puzzles and Paradoxes in Statistics, First Edition.
Eric Sowey and Peter Petocz.
© 2017 John Wiley & Sons, Ltd. Published 2017 by John Wiley & Sons, Ltd.
Companion website: www.wiley.com/go/sowey/apanoramaofstatistics

the single measurement stated is simply to be taken to be the correct one, and the assigned task goes on from there.

If we ask why 100 people don't all produce exactly the same length measurement for the same object, using the same tape measure, we are led to a fundamental realisation. There are many factors at work in the physical act of measurement that cause different results to be reported for the same task by different people. Among these factors are: the attentiveness with which the task is undertaken; the effect of parallax in reading the scale marks on the tape; possible tremor in the hand holding the tape measure; and eyesight variations in those reporting measurements. Some of these factors might lead a person to mismeasure in a way that exceeds the correct value, while other factors might cause that same person to fall short of the correct value. Moreover, different people might react differently to any particular factor.

While it would be theoretically possible to study systematically some, or all, of these causes of variation individually (and there are contexts where it would be important to do so), it is generally convenient to lump all these real-world factors together and to refer to their net effect on measurement as *chance* (or *random*) *variation* around the correct value. This highlights the truth that chance influences are inseparable from almost all experience of life in the real world. (For more detail about the meaning of randomness, see CHAPTERS 10 and 11.)

Chance variation has long been recognised. A famous passage in the biblical Book of Ecclesiastes, written some 2,200 years ago, shows how random events can have perplexing impacts: '… the race is not to the swift, nor the battle to the strong, nor bread to the wise, nor riches to the intelligent, nor favour to those with knowledge, but time and chance happen to them all.'

One may, of course, choose to abstract from chance influences (as the Public Transport Department does, for example, when it publishes a train timetable), but looking away from them should be understood as a deliberate act to simplify complex reality. In contexts where chance effects are ordinarily small (e.g. train journey times along a standard urban route), abstracting from chance is unlikely to cause decision errors to be made frequently (e.g. about when to come to the station to catch the train). However, where events are heavily dominated by random 'shocks' (e.g. daily movements in the dollar/pound exchange rate on international currency markets), predictions of what will happen even a day ahead will be highly unreliable most of the time.

As we mentioned in CHAPTER 1, school mathematical problems are generally posed in abstract settings of complete certainty. If, on occasion, a mathematical problem is posed in an ostensibly real-life setting, the student is nevertheless expected to abstract from all chance influences, *whether doing so is true to life or not*. Here is a typical example: try solving it now.

According to the timetable, a container train is allowed 90 minutes for a journey of 60 miles over mountainous country. On a particular trip, the train runs into fog and its speed is reduced, making it 25 minutes late at its destination. Had the fog appeared six miles closer to its point of departure, the train would have been 40 minutes late. At what rate does the train travel through fog?

(The answer is 15 miles per hour. Did you see in which way the problem is unrealistic and where chance influences are ignored? The train, incidentally, is pictured on the cover of this book.)

However appealing such problems may be for exhibiting the 'relevance' of mathematics, they risk concealing from students its fundamental logical limitation. The great physicist, Albert Einstein (1879–1955), expressed it concisely: 'As far as the propositions of mathematics refer to reality, they are not certain; and as far as they are certain, they do not refer to reality.' (see Einstein (1921), online in the German original at [2.1] and in English translation at [2.2]).

We can elaborate Einstein's aphorism like this. In solving a problem in a real-life setting, the mathematical approach neglects all chance influences in that setting and, *on that account*, the mathematical solution is stated with certainty – but that solution is evidently an approximation to the solution in reality. Moreover, the error in the approximation is indeterminate. The statistical approach, by contrast, recognises the chance influences explicitly and, *on that account*, the statistical solution cannot be stated with certainty. The statistical solution, too, is an approximation to the solution in reality – but in the statistical approach, the error due to the chance influences *can* be dependably assessed within bounds.

Well, what about problems in probability? Self-evidently, they are problems about chance events but, here, calculating the probability of occurrence of some random event is the entire goal: it is simply an exercise in arithmetic according to predefined rules. Moreover, within the scope of the problem, it is certain that the calculated probability is correct. Therefore such problems, too, are mathematical problems. However, were the random event embedded in some inferential context, then the inferential problem would *thereby* be a statistical problem.

---oOo---

So far, we have seen that the central focus of statistics is on variation and, in particular, on chance variation. Mathematics acknowledges variables, but it does not focus *centrally* on their variation, and it abstracts entirely from the influence of chance.

The central focus of mathematics is on those general properties that are common to all the varying members of a set. Circles, for example, vary in their diameters, but circle theorems relate to the properties that all circles have in common, regardless of their particular diameter. Similarly, Pythagoras's Theorem is true for all plane right-angled triangles, regardless of their size.

While mathematicians' prime objective is to prove general theorems, which then imply truths about particular cases, statisticians proceed in reverse. They start with the 'particular' (namely, a sample of data) and, from there, they seek to make statements about the 'general' (that is, the population from which their data were sampled).

---oOo---

Finally, we come to the contrasting nature of numerical data in mathematical and in statistical problems. Data (literally, 'givens' – from the Latin) are indispensable inputs to any process of computational problem solving. However, 'data' mean different things to mathematicians and to statisticians.

As we have seen, to a mathematician data are values of non-random variables, and the task is to apply those numbers to evaluate a special case of a known general theorem – for example, to find the equation of the (*only*) straight line that passes (*exactly*) through two points with given coordinates on a plane. To a statistician data are values of random variables, and the statistician asks, 'How can I confidently identify the underlying systematic information that I think there is in these data, but that is obscured by the random variability?' For example, what is the equation of the *best-fitting* straight line that passes *as near as possible* to ten points with given coordinates, scattered about on the same plane in a pattern that looks roughly linear, and how well does that scatter fit to that line? The statistician also asks, 'How reliably can I generalise the systematic sample information to the larger population?' A large part of a practising statistician's work is the analysis of data – but a mathematician would never describe his or her work in this way.

As if practising statisticians did not have enough of a challenge in seeking out meaningful systematic information 'hidden' in their randomly-varying data, they must also be prepared to cope with a variety of problems of data quality – problems that could easily send their analysis in the wrong direction.

Among such problems are *conceptual errors* (a poor match between the definition of an abstract concept and the way it is measured in practice), *information errors* (e.g. missing data, or false information supplied by survey respondents), and *data processing errors* (e.g. the digital transposition that records the value 712 as 172).

What follows from this perspective on the ways in which statistics differs from mathematics? Statistics is not independent of mathematics, since all its analytical techniques (e.g. for summarising data and for making statistical inferences) are mathematically derived tools (using algebra, calculus, etc). However, in its prime focus on data analysis and on generalisation in contexts of uncertainty, and in its inductive mode of reasoning (described in CHAPTER 1), statistics stands on its own.

Questions

Question 2.1 (A)

a) Pure mathematics deals with abstractions. In geometry, for example, lines have length but no breadth, planes have length and breadth but no thickness, and so on. In this setting, we pose the problem: a square has sides 20 cm long; what is the length of its diagonal? How does the mathematician answer? Is this answer accurate?

b) Statistics deals with real-world data – measurements of actual objects and observations of actual phenomena. On graph paper, construct, as precisely as you can, a square with sides 20 cm long. Then ask, say, 25 different people to use their own millimetre-scale rulers to measure the length of the square's diagonal, taking care to be as accurate as possible. Record these measurements. In this setting, we pose the problem: what is the length of the diagonal? How does the statistician answer? Is this answer accurate?

Question 2.2 (A)

Mathematical induction is a versatile procedure for constructing a proof in mathematics. Explain the general approach underlying mathematical induction, and give an example of its use. Does mathematical induction use deductive logic or inductive logic? What is the implication of your answer to this question?

Question 2.3 (A)

Consider the mathematical expression $n^2 + n + 41$. If $n = 0$, the expression has the value 41, which is a prime number. If $n = 1$, the expression has the value 43, also a prime number.

a) Make further substitutions of $n = 2$, 3, 4, 5 and, in each case, check whether the result is a prime number. What does this suggest?

b) Repeat for $n = 6, \ldots, 10$ (or further, if you like). What inference would you draw statistically from this accumulating information?

c) Is this inference actually correct? Can you prove or disprove it, mathematically?

Question 2.4 (A)

Sherlock Holmes, the famous fictional consulting detective created by the British novelist Sir Arthur Conan Doyle (1859–1930), solved crimes by reasoning from data in ways similar to a statistician's reasoning. What similarities can you see in the following passage (from 'The Five Orange Pips' in the book *The Adventures of Sherlock Holmes*)?

> 'Sherlock Holmes closed his eyes, and placed his elbows upon the arms of his chair, with his fingertips together. "… [W]e may start with a strong presumption that Colonel Openshaw had some very strong reason for leaving America. Men at his time of life do not change all their habits, and exchange willingly the charming climate of Florida for the lonely life of an English provincial town. His extreme love of solitude in England suggests the idea that he was in fear of someone or something, so we may assume as a working hypothesis that it was fear of someone or something which drove him from America. As to what it was he feared, we can only deduce that by considering the formidable letters which were received by himself and his successors. Did you remark the postmarks of those letters?" "The first was from Pondicherry, the second from Dundee, and the third from London" [replies Dr Watson]. "From East London. What do you deduce from that?" "They are all seaports. That the writer was on board a ship." "Excellent. We have already a clue. There can be no doubt that the probability – the strong probability – is that the writer was on board of a ship."'

Why is the word 'deduce' inappropriate in the above passage? What word should Conan Doyle have used instead?

Question 2.5 (A)

How can one find an approximation to the value of π, the ratio of a circle's circumference to its diameter, by way of a statistical experiment that involves tossing a needle randomly onto a flat surface ruled with parallel lines? What is the name of the 18th century polymath with whom this experiment is associated?

References

Online

[2.1] Einstein, A. (1921), *Geometrie und Erfahrung.* Erweiterte Fassung des Festvortrages gehalten an der Preussischen Akademie der Wissenschaften zu Berlin am 27. Januar 1921. Julius Springer, Berlin. Pages 3–4. At https://archive.org/details/geometrieunderf00einsgoog

[2.2] Einstein, A. (1921), *Geometry and Experience.* Lecture before the Prussian Academy of Sciences, January 27, 1921. Julius Springer, Berlin. In The Collected Papers of Albert Einstein. At http://einsteinpapers. press.princeton.edu/vol7-trans/225

3

Statistical literacy – essential in the 21st century!

Numeracy, also called quantitative literacy, is an undervalued skill that struggles for recognition among educators and the general public. It is only since the turn of the century that governments in many developed countries have come around to recognising that a numerate citizenry is as important an attribute of an advanced society as a literate citizenry. But there have been lone voices urging this recognition for a long time, among them John Allen Paulos, whose 1988 book *Innumeracy: Mathematical Illiteracy and its Consequences* made a considerable impact. Already more than a century ago, the British writer H.G. Wells foresaw that advancing numeracy would inevitably become a pressing need for society (for more on Wells' prediction, see QUESTION 5.4).

By 'numerate', here, we mean 'functionally numerate'. A basically numerate person is someone who recognises number symbols and correctly performs basic arithmetic. A functionally numerate person can also correctly interpret and meaningfully evaluate a logical argument couched in numbers. It should be clear that one does not need to know any advanced mathematics to be functionally numerate.

However, what *is* needed is not what one finds emphasised in calls for reform of the traditional high school mathematics curriculum. That is to say, making traditional school mathematics topics more interesting, more relevant to a career such as engineering or accountancy, or more fun, will not necessarily produce highly numerate adults. A curriculum for effectively developing functional numeracy looks rather different from a curriculum, however thoughtfully it is enhanced, for teaching algebra, calculus, geometry and trigonometry.

A Panorama of Statistics: Perspectives, Puzzles and Paradoxes in Statistics, First Edition.
Eric Sowey and Peter Petocz.
© 2017 John Wiley & Sons, Ltd. Published 2017 by John Wiley & Sons, Ltd.
Companion website: www.wiley.com/go/sowey/apanoramaofstatistics

As we shall explain, it is topics in statistics, rather than topics in mathematics, that lie at the heart of any effective movement to advance quantitative literacy in the community. *Quantitative literacy is, in essence, statistical literacy.*

There were quantitative literacy movements even when government backing was hardly yet available. They developed out of the creative efforts of individuals and scholarly societies, who understood how essential it is in a liberal society to equip citizens with the knowledge to dissect the kinds of quantitative arguments increasingly levelled at them – in particular, by politicians and marketers set on persuasion. Among the consistently active advocates for a numerate society over the past 20 years have been Andrew Ehrenberg (in the UK) and Lynn Arthur Steen (in the USA).

Current numeracy projects are still mostly evolving within national boundaries, though some cross-fertilisation is emerging as a result of initiatives by scholarly societies, and of interactions nurtured at international conferences.

At the school level, a particularly promising UK initiative that has now taken on an international dimension is a project called CensusAtSchool, a 'children's census' aiming to collect and disseminate data about, and relevant to, students. CensusAtSchool started in 2000, with support from the Royal Statistical Society. It was initially run by the International Centre for Statistical Education at Plymouth University in the UK. Despite recent widespread funding cuts, several countries continue to participate. The website at [3.1] gives details of the UK project. The Irish site is at [3.2] and the Canadian site at [3.3]. School teachers and their students have access to data from their own and other countries. These data provide a rich context in which to develop a basic understanding of the uses and limitations of numerical information, as a practical introduction to statistical literacy.

An international overview of projects and publications in statistical literacy is maintained by the International Statistical Literacy Project, which is online at [3.4]. The ISLP operates under the auspices of the International Statistical Institute.

Another valuable website on this theme, at [3.5], is maintained in the USA by Milo Schield, who is committed to giving a lead on enhancing people's statistical literacy and is, himself, a prolific writer on the subject. Schield's website is a comprehensive source of information, aimed particularly at adults. The home page has links to a large number of articles and books – popular as well as academic – on various aspects of statistical literacy. There is enough reading there to keep anyone busy for a long time.

We think it appropriate that current approaches to advancing quantitative literacy in schools are focusing on statistical literacy, rather than on mathematical literacy. As we point out in CHAPTER 1, mathematical methods are not routinely applicable to interpreting and evaluating quantitative arguments about the real world, where the influence of uncertainty is generally inescapable. However, that is precisely the realm of statistics. A statistical approach to promoting adult numeracy appeals because it can embrace discussions of practical issues in society, such as the broad topic of social justice (see Lesser (2007) – online at [3.6]). Further, such discussions often spark new interest in quantitative matters among people who studied only (deterministic) mathematics in their schooldays, and puzzled at that time over the apparent disconnection between textbook problems and reality.

Together with these developments in school and adult education, there is a parallel growth of academic research interest in the area of statistical literacy. A comprehensive collection of publications in statistics education between 2010 and 2014 was studied by one of the present authors (PP). It revealed that 60% of articles referred to at least one of the terms 'statistical literacy', 'statistical reasoning' and 'statistical thinking'. Since 1999, a group of scholars has been actively investigating these three topics. Some of the group's activities are outlined on their SRTL (Statistical Reasoning, Thinking and Literacy) website at [3.7].

Our QUESTIONS 3.1 and 3.2, below, are based on items in the media. These questions hint at what a handicap statistical illiteracy can be to a competent understanding of public discussions on topics of social and community interest.

Questions

Question 3.1 (B)

A report on the *BBC News* (11 February 2005 – online at [3.8]), related the story of the number 53 in the Venice lottery. The Italian national lottery is a type of lotto, in which draws of five numbers from numbers 1 to 90 take place in each of ten cities across the country (a total of 50 numbers are selected). In the Venice lottery, the number 53 had not come up in 152 consecutive draws over 20 months, and Italians were in a frenzy betting on 53 in Venice, ever more utterly convinced on each occasion that it failed to appear, that it simply had to appear next time. Four people reportedly died, and many others were completely ruined in incidents related to this betting.

a) What name is popularly given to the (invalid) principle which motivated the public's conviction that 53 was ever more likely to appear, the longer the continuous run of its non-appearance?

b) What is the probability that number 53 does not come up 152 times in a row in the Venice lottery? What is the probability that it does not come up 152 times and then does come up on the 153rd draw (as actually happened)? In this context, where there are 90 numbers to choose from at each draw, what related lottery outcome might also have caused a betting frenzy, even though the probability of its occurrence is much larger?

Question 3.2 (A)

To judge from the frequency with which they turn up in news reports of social research, statistical 'league tables' have a strong appeal to the public – perhaps because they give each social group the satisfaction of seeing how many other groups they are 'outperforming'. But such league tables can be a rich source of misconceptions.

FIGURE 3.1 shows data on the 'Top 10 night-owl nations' from an international survey of sleep hours in 28 countries in Europe and the Asia-Pacific region, carried out in 2004 by a reputable agency. The survey procedure is described as follows: 'a representative sample of consumers with internet access are asked questions relating to … their attitudes towards a variety of topics.' On this occasion about 14,000 people were surveyed.

Conclusions are presented in the following terms: 'the biggest night-owls the world over are the Portuguese, with 75% not "hitting the sack" until after midnight … The second ranked global night-owls are the Taiwanese, with 69% going to bed after midnight … Following closely behind are the Koreans (68%) …'

Rank	Country	Going to bed after midnight
1	Portugal	75%
2	Taiwan	69%
3	Korea	68%
4	Hong Kong	66%
5	Spain	65%
6	Japan	60%
7	Singapore	54%
8	Malaysia	54%
9	Thailand	43%
10	Italy	39%

Figure 3.1 Bedtime habits by country.

What critical questions should occur to a statistically literate person, regarding this 'league table' and the survey information and conclusions, as described?

Question 3.3 (A)

You are asked a question by a beginner in statistics: 'For valid inferences about a diverse population, why isn't it more important to have a representative sample than a random sample?' How do you reply?

Question 3.4 (A)

It is well known that the shape of a histogram will vary, depending on the number of class intervals used in its construction. What other features of its construction influence the shape of a histogram? Is there a 'best' choice for the number of class intervals to use in practice?

Question 3.5 (A)

The histogram in FIGURE 3.2 shows the ages of 282 of the 778 convicts who were transported on the *First Fleet* from England to Australia in 1788. Can you use this information to estimate the mean age of all the convicts on the *First Fleet*? If you can, what is the value of your estimate?

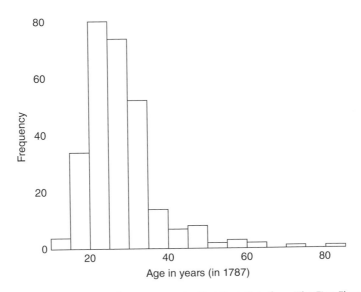

Figure 3.2 Ages of a sample of convicts on the *First Fleet*. Data from 'The First Fleet Convict Database', Computer Education Unit, NSW Dept. of Education, 1986.

References

Print

Paulos, J.A. (1988). *Innumeracy: Mathematical Illiteracy and its Consequences.* Penguin.

Online

[3.1] http://www.censusatschool.com/
[3.2] http://www.censusatschool.ie/
[3.3] http://www.censusatschool.ca/
[3.4] http://iase-web.org/islp/
[3.5] http://www.statlit.org
[3.6] Lesser, L.M. (2007). Critical values and transforming data: teaching statistics with social justice. *Journal of Statistics Education* 15 (1). At http://www.amstat.org/publications/jse/v15n1/lesser.html
[3.7] http://srtl.fos.auckland.ac.nz/
[3.8] http://news.bbc.co.uk/2/hi/europe/4256595.stm

4

Statistical inquiry on the web

As a discipline serving other disciplines, statistics has been ubiquitous in print for half a century, and its applications can be found reported in almost every field of human endeavour. Since 2000, statistics also appears widely across the web, in both closed-access and open-access documents.

It is relevant here to distinguish these two classes of web documents. The former (located behind some form of pay-wall) are mostly documents produced by commercial organisations (book and journal publishers, software designers, and consultancies of every kind). However, items that have traditionally been closed-access are increasingly being made partially available in open-access form. This is partly due to the encouragement of major online retailers and search engine companies, who propose that tempting readers with free access to a strictly limited number of document pages can elicit increased commercial sales in the long run. There is also a vigorous grass-roots movement worldwide, pressing for the transformation of closed-access scholarly journals to open-access. Now that there are several models for publishers to recoup, in other ways, the income they would forgo by abandoning access charging, that transformation is gaining pace.

In what follows, we shall write only about the open-access web.

It is quite impossible to give an overview of all the material on, or related to, statistics – in all its senses: the discipline, the techniques, and the numerical data – that is currently on the web in English and other languages. However, some idea of the variety and scope of this material can be obtained by browsing some of the leading English language websites devoted to bringing together links to online statistical material of high quality.

Wide-ranging, and for statisticians generally, is Gordon Smyth's portal at [4.1]. A parallel site focused on probability is *The Probability Web* at [4.2].

A Panorama of Statistics: Perspectives, Puzzles and Paradoxes in Statistics, First Edition.
Eric Sowey and Peter Petocz.
© 2017 John Wiley & Sons, Ltd. Published 2017 by John Wiley & Sons, Ltd.
Companion website: www.wiley.com/go/sowey/apanoramaofstatistics

A detailed reference manual of techniques is Gerald Dallal's *The Little Handbook of Statistical Practice* at [4.3]. Two extensive online textbooks are Keith Dear's *SurfStat* at [4.4] and David Lane's *HyperStat* at [4.5]. The fertility of statistical methods in a wide variety of fields is conveyed by the collection of videoed lectures on the *Chance* website at [4.6].

The field of statistics education is well served by the compendium of links on the US CAUSEweb organisation's site at [4.7], and by the open-access academic journals, *Journal of Statistics Education* at [4.8] and *Statistics Education Research Journal* at [4.9]. Full-text access to hundreds of other academic journal articles on statistics is available via the Project Euclid searchable database at [4.10]. A portal to every country's official national agency that collects and publishes demographic and economic statistics is linked at [4.11]. You can find several websites providing collections of links on the history of probability and statistics in CHAPTER 22.

As everyone knows, anything to do with the web changes fast, sometimes very fast! The rate at which new material appears is such that, for those who maintain sites like those just mentioned, keeping up is an enormous task. Even the most dedicated and persevering individual is eventually overwhelmed. Regrettably, few of the online information repositories about statistics are assured continuity of staffing (not to speak of funding).

At the same time, countless documents are deleted from the web or have their uniform resource locator (URL) address changed without providing a forwarding link. That means that their links in online information repositories become 'dead'. Dead links, especially those to valued and frequently accessed resources, are naturally a source of frustration to all.

To reduce the incidence of dead links among popular websites, it would be good to have a way of identifying a web document as 'stable' – that is, destined for permanent residence on the web, and automatically traceable by a search engine even if its URL address is changed. Several systems for stable document identification are, indeed, currently available. The most widely adopted such system in the academic sphere is, at present, the Digital Object Identifier (doi), described online at [4.12]. However, since it is a fee-based product, it is uncommon to find open-access documents so identified.

---oOo---

What with the continual addition and deletion of web documents, two successive searches using the same search terms – even just a day apart – may produce a different number of hits. How much more marked would this effect be if the interval between the searches were far longer? To explore this, we present a little case study that took us into a literary byway of statistics. It contrasts what we found in a specific search in 2007 with exactly the same search done eight years later.

One of us remembered reading, in pre-web days, some statistician's claim to the effect that 'we may not be able to use statistics to answer the question of what song the Sirens sang, but it can be used to answer just about anything else'. Might a web search enable us to rediscover who made that claim?

The Sirens, we remind you, were fabled creatures in Greek mythology who, from their island seat high on a rock beside the water, could lure sailors to their destruction on a reef below by the seductiveness of their singing. Statistical mariners were, however, distracted differently, it seems.

'Don't bother singing to this lot, girls. Just show them your figures.'

Using Google to search for the exact words 'song the Sirens sang' ultimately (i.e. after progressing to the last of the search result screens) generated about 80 hits. Among these we found two remarkable items. Firstly, the question 'What song did the Sirens sing?' was (so the Roman historian Suetonius reported) asked teasingly of visiting scholars of Greek by the Roman emperor Tiberius some 2000 years ago. This question has come down to us today as symbolic of all legendary enigmas (though what Tiberius was probably getting at was that the Sirens, usually portrayed in that era as part-woman and part-bird, were more likely to have been warbling alluringly rather than singing *any* song).

Secondly, Sir Thomas Browne, an eminent doctor and writer of Norwich, England, observed cautiously in 1658, 'What song the Sirens sang, or what name Achilles assumed when he hid himself among women, though puzzling questions, are not beyond all conjecture.' Thus, we discovered that the 'song the Sirens sang' is a much older allusion than we had believed.

Continuing with the additional search term 'statistics' narrowed the field to a mere 20 hits. None of these, alas, yielded a phrase resembling the quotation whose author we were seeking.

Eight years later, we again made a Google search for the exact words 'song the Sirens sang'. This time we scored about 210 hits. Adding the search term 'statistics' reduced the number of hits to about 70 (we say 'about' because, as already mentioned, web search hits fluctuate somewhat by the day). Once again, however, there was no success in identifying the author of the remembered quotation.

Nonetheless, this little experiment has revealed or confirmed some interesting things.

Firstly, we note that, with the search terms unchanged, the number of hits increased substantially over the eight years in both searches – from 80 and 20 to 210 and 70. This is hardly surprising, given the enormous annual growth rate of the internet. Looking specifically at the hits found in our particular searches lets us identify three prominent forces behind the growth in their numbers. These forces are the growth of blogging, the growth in numbers of websites with English language content originating from non-Anglophone countries, and the growth in numbers of print publications migrating – at least partially – to the open-access web.

Next, we look at the quality of the harvested results relating to statistics. About 20% of the hits in our most recent search, coupling Sirens and statistics, were on documents dealing directly with statistical theory or applications (as opposed to literary speculations about the legendary Sirens). Most of these hits were either scholarly items or educational materials. Questions of the accuracy, meaningfulness and quality of explanation in these documents arise – as they do, of course, in regard to all information gleaned from the web. Except where the material has had some form of impartial peer review or has other scholarly standing, scepticism may be the most appropriate initial approach to the information offered.

Further, the web hosts many publicly-modifiable wiki-type resources dealing with statistical topics, as well as websites where anyone can 'vote' (anonymously) for the 'best' of several (anonymously) submitted answers to someone's uploaded statistical problem. Our search turned up one such site. We can only reiterate the routine warnings of statistics professionals to treat the information on these sites with extreme caution.

---oOo---

And now, what about the half-remembered quotation from a statistician about the almost universal utility of statistics? In a serendipitous moment while this chapter was in draft, the source came to mind, after half a lifetime, in a sudden flash of recollection. It was the first edition of Freedman, Pisani and Purves (1978), where the cautious words of Sir Thomas Browne (as already cited) appeared in the Preface. So it was not *one* statistician who claimed that statistics is almost universally useful, but *three*. Enlivened by Browne's words, these authors declare 'Statistics is the art of making numerical conjectures about puzzling questions' (page xiii), inviting the reader to understand that, though there *may* be questions about the world whose answers are ultimately unknowable, answers to the questions with which statistics is equipped to deal are, indeed, 'not beyond all conjecture'.

The quotation from Sir Thomas Browne is a feature of all editions of the text by Freedman, Pisani and Purves, the latest having been published in 2007. Why, then, was this location of Browne's words not one of the hits of our web searches? Even though books in copyright are rarely on the open-access web, many publishers (as mentioned earlier) see commercial merit in making their books partially open access. The Freedman, Pisani and Purves book is an example. Unfortunately, the page bearing the quotation turns out not to be among the open-access pages.

Questions

Question 4.1 (A)

Which seventeenth century Italian physician and inveterate gambler wrote the following (given in translation), and in which book?

'Even if gambling were altogether an evil, still, on account of the many people who play, it would seem to be a natural evil. For that reason, it ought to be discussed by a medical doctor like one of the incurable diseases.'

Question 4.2 (A)

In what theatre play do the protagonists toss coins repeatedly and obtain 100 heads in a row? What do they philosophise about this occurrence?

Question 4.3 (A)

When was the earliest official census in England undertaken and how was it documented?

Question 4.4 (B)

In 1935, a certain Dr Anderson, a botanist, wrote an article about a particular kind of flower. A year later, his botanical measurements were applied by a famous statistician to illustrate the development of a new statistical technique.

   What was the flower and where were the data collected? Who was the famous statistician and what was the new statistical technique being illustrated?

Question 4.5 (B)

In the mid-1700s, a famous Italian adventurer persuaded the French government, with the help of the mathematician and philosopher Jean D'Alembert, to set up and run a government-backed lottery to raise money for a military academy in Paris. Who was the adventurer? How long did the lottery survive? Although it was an early version of our contemporary lotto, there was an important difference – what was this?

References

Print

Freedman, D., Pisani, R. and Purves, R. (1978). *Statistics*, 1st edition, Norton.

Online

[4.1] http://www.statsci.org
[4.2] http://probweb.berkeley.edu/probweb.html
[4.3] http://www.StatisticalPractice.com
[4.4] http://surfstat.anu.edu.au/surfstat-home/surfstat.html
[4.5] http://davidmlane.com/hyperstat
[4.6] http://www.dartmouth.edu/~chance/ChanceLecture/AudioVideo.html
[4.7] https://www.causeweb.org/resources
[4.8] https://www.amstat.org/publications/jse
[4.9] http://iase-web.org/Publications.php?p=SERJ
[4.10] https://projecteuclid.org/
[4.11] http://www.unece.org/stats/links.html#NSO
[4.12] http://en.wikipedia.org/wiki/Digital_object_identifier

Part II

Statistical description

5

Trustworthy statistics are accurate, meaningful and relevant

Statistics, as the term appears in this chapter's title, are numbers relating to the real world that are collected systematically for a purpose. Have you noticed how often statistics are mentioned these days in the print and digital media, as well as on websites of every kind? These statistics are usually precise, and it is usually implied that they are accurate. We remind you that precision (i.e. exactness) and accuracy (i.e. correctness) are not the same thing.

Among headlined statistics, some are counts ('world population passes seven billion'), some are averages ('Australia's population density among the world's lowest'), some are league tables ('the annual *Rich List*'), and some are estimates generalised from sample data ('84% of us dislike eating offal'). These statistics serve many different purposes – to inform, to impress, to persuade and, sometimes, just to entertain. There is even an unthinking

Numbers you *do* need.

A Panorama of Statistics: Perspectives, Puzzles and Paradoxes in Statistics, First Edition.
Eric Sowey and Peter Petocz.
© 2017 John Wiley & Sons, Ltd. Published 2017 by John Wiley & Sons, Ltd.
Companion website: www.wiley.com/go/sowey/apanoramaofstatistics

fashion in some newspapers and magazines to embellish news items or feature articles on 'boring' subjects with a text box displaying supposedly related statistics ('the numbers you need'), but arranged in no particular order, and adding negligible value to the story. They are, in fact, numbers you *don't* need!

This rain of numbers may be journalists' over-enthusiastic response to discovering that people seem to be fascinated by numerical statistics when they illuminate the context – *and even when they don't*!

Unfortunately, because the general level of numeracy (also called 'statistical literacy') in the community is quite low, even in countries with comprehensive education systems, misinterpretations of statistics and misunderstandings of statistical methods are very common. This, then, opens a door to those who want to misuse statistics in order to mislead or to deceive. In CHAPTERS 8 and 9, we review some of the ways in which statistics can be misleading.

Happily, increasing efforts are now being made in many countries to remedy the general public's limited understanding of basic statistical ideas and their lack of experience in interpreting numerical statistics. Notable among these efforts are the active statistical literacy initiatives of the Royal Statistical Society, which can be found via the RSS website, online at [5.1], and the evolving International Statistical Literacy Project, online at [5.2].

For an inexperienced consumer of statistics, it can be difficult to discriminate trustworthy (i.e. accurate, meaningful and relevant) statistics from those that may be unreliable (at best) or deceptive (at worst). To come to a decision on this issue, one needs answers to these questions:

- Who produced these statistics? (the maker)
- How were they produced? (the method)
- Who is presenting them to me, and why? (the motive)

Asking about the *maker* and the *method* is a way of judging how accurate and meaningful the statistics are. Introspecting about the *motive* can prompt you to assess the relevance of statistics that are being used persuasively in support of a particular cause, and to help you to decide whether the declared cause is, in fact, the *real* cause.

Thus, the answers you get to these questions can not only identify untrustworthy aspects of the statistics, but also greatly enrich your understanding of what *is* trustworthy and why.

---oOo---

To illustrate, we shall look at three fundamentally important statistics – one global, one national, and one personal. The statistics are values of the following variables: the concentration of carbon dioxide in the atmosphere; the national unemployment rate; and a person's white blood cell count.

Whatever the source of his or her information, when a statistically literate person comes upon a value for any of these variables, he or she will want to know if it is accurate, meaningful and relevant.

As we indicated above, finding out about accuracy and meaningfulness is best done by learning something about the maker of the statistic and the method of measurement. Our experience is that these inquiries often turn up unexpectedly interesting information. About relevance, we have something to say at the end of this Overview.

(a) The concentration of carbon dioxide in the atmosphere

Where might we come across this statistic? Perhaps in a magazine article about environmental protection that announces that the concentration by volume of carbon dioxide (CO_2) in the Earth's atmosphere is approaching 400 parts per million.

Here is some background. The Earth's atmosphere is currently composed of 78.08% nitrogen, 20.94% oxygen, 0.93% argon and 0.04% CO_2, by volume. The remaining volume (less than 0.01%) is a mixture of about a dozen other gases and water vapour, each present in minute quantities. Unlike nitrogen, oxygen and argon, CO_2 is a 'greenhouse gas' – that is, it has the property of absorbing some of the sun's heat that is reflected from the Earth's surface and radiating it back to Earth. So, although it comprises only such a tiny part of the atmosphere, CO_2 plays a large part in determining the ambient temperature on the Earth's surface. Even a quite minor rate of increase in the concentration of atmospheric CO_2 could, if it were sustained, have major long-term consequences for the planet, as a result of 'global warming'.

Now to the *makers* (i.e. the data collection and processing agencies) and the *methods* of measurement – the two factors that are central to judging the accuracy and meaningfulness of the statistics. We leave it to you to make these judgements by following up the references in the next two paragraphs, together with others you may discover for yourself.

Detecting any change in the tiny atmospheric concentration of CO_2 evidently needs very accurate measurement. Nowadays, high accuracy is achieved using spectrometers mounted on aircraft or satellites. These instruments measure changes in sunlight as it passes through the atmosphere. Good results can also be obtained spectrometrically by analysing air samples captured at ground level, whether on land or at sea. To ensure that the measurement is meaningful, it makes sense to select places where atmospheric gases are well mixed – that is, away from locations dense with CO_2 generators, whether from human activity (e.g. burning of fossil fuels) or natural (e.g. the rotting of vegetable matter). A good overview of many current worldwide sites, their respective CO_2 measurement procedures, and the data they have yielded, can be found on the website of the Carbon Dioxide Information Analysis Center at Oak Ridge National Laboratory, USA, online at [5.3]. A similar overview site is operated by the World Data Centre for Greenhouse Gases, Japan, online at [5.4].

To construct a monthly indicator for each data collection site, a daily average is calculated from hourly data, and a monthly average from averaging the daily averages. In this way, it is hoped to damp the influence of outliers in the initial measurements. Subsequently, monthly indicators from many locations are combined. A glimpse of how measurements from multiple sites are combined by complex methods of curve-fitting can be obtained via the live link, labelled 'more details on how global means are calculated', on the webpage of the Earth System Research Laboratory (ESRL), USA, online at [5.5].

Ultimately, a consensus monthly figure for global atmospheric CO_2 concentration is settled upon, across multiple data collection agencies and multiple measurement sites. At the time of writing (in April 2016), this figure from the ESRL is 404.08 parts per million. Note that this is a point estimate. The ESRL provides an indication of uncertainty with a bootstrap-estimated standard deviation. Currently, this is 0.09 parts per million.

(b) The national unemployment rate (NUR)

Where might we come across this statistic? For example, in a newspaper report stating that Australia's seasonally-adjusted national unemployment rate (NUR) is currently 5.9%.

Here is some background. In Australia, the value of the NUR (i.e. the percentage of the labour force that is unemployed) is calculated by the Australian Bureau of Statistics (ABS, the national statistics office) from sample data that it collects around the country. Labour force statistics, including the value of the NUR, are given in the ABS monthly print publication 6202.0, *Labour Force, Australia*. This publication can also be downloaded (without charge) in page-image form from the ABS website at [5.6].

Explanatory notes at the end of the publication (and also at the Explanatory Notes tab of the web page from which the publication can be downloaded) describe in some detail how the raw NUR figure is constructed, and how its seasonally adjusted value is derived. (Seasonal adjustment makes allowance for short-term fluctuations in unemployment that are 'built-in' to the economic cycle over the year. For example, unemployment always rises for a month or two after the end of the academic year, when thousands of secondary and tertiary students go looking for a job. It always falls around the time of major religio-commercial festivals, such as Christmas.) There is also a 95% confidence interval for the true value of the NUR.

These many explanations and clarifications help statistics users to gauge the accuracy and meaningfulness of the ABS-calculated value of this widely cited economic indicator.

You may be able to find out how the official NUR is determined in your own country from the website of your national statistics office. Links to the websites of these offices can be found at [5.7].

(c) The white cell count

Where might we come across this statistic? In a patient's blood test report, sent to his/her doctor, showing the actual white blood cell count, together with the annotation that the 'reference range' is 3.8 to 10.8.

Here is some background. The white blood cell count is expressed in thousands of cells per microlitre of blood. A microlitre is equivalent to a cubic millimetre. A person's white cell count is determined from a sample of blood, generally by a registered pathology laboratory, using specialised equipment. The accuracy of the count depends on the quality of the equipment and the competence of the technologist. In Australia, these are required to be reassessed regularly (commonly, at two-yearly intervals) prior to official renewal of registration.

A white cell count that falls within the reference range is described as 'normal', in the sense that no disease that markedly elevates or depresses the white cell count (relative to that of a healthy person) is likely to be present. However, even in a registered laboratory, it can happen that a white cell count is inaccurate. Then, the cell count might be a 'false positive' – signalling that disease is present when, in fact, it is not. Or it might be a 'false negative' – signalling that disease is not present when, in fact, it is. Of course, one hopes that such inaccuracies are rare.

What about meaningfulness? The meaningfulness of the white cell count as a diagnostic aid depends substantially on the reliability of the reference range as a demarcation criterion between what is normal and what is abnormal.

You may be surprised to learn that the limits of the reference range are not internationally agreed (in contrast, for example, to the world standards for units of length, mass and time). Rather, these limits are the endpoints of a 95% statistical confidence interval *constructed by each pathology laboratory from its own records* of the vast number of white cell counts it has performed. An approximate 95% confidence interval is given by the mean of all the laboratory's recorded white cell counts, plus or minus twice the estimated standard error of the mean. It follows that slightly different reference ranges can be expected in reports from different laboratories. Indeed, our own experience over the years includes the following ranges: (3.5–11.5), (3.7–11.4) and (4–11).

This is one sign that the reference range has its limitations. Here are two further signs. The reference range is calculated using data from a mix of healthy and unhealthy individuals. It would, clearly, not be useful as an indicator of the 'normal range' if the proportion of the data coming from unhealthy individuals was high. Further, the 'normal range' of white cell counts differs between infants and all other individuals.

--oOo---

So far, we have considered issues of accuracy and meaningfulness in numerical statistics. Ensuring that a quoted statistic is also relevant to its context – say, a public policy debate on some controversial matter – is the third aspect of the trustworthiness of a statistic. It is in this connection that one must inquire into the presenter's *motive* in quoting the statistic.

If the presenter is disinterested in (i.e. neutral about) the topic of debate, then an ulterior motive is unlikely. But a disinterested presenter is rarely encountered when the matter is controversial. Thus, it is important to check: on the source of the statistic; on whether the presenter has quoted the statistic correctly; on whether any weaknesses of the statistic have been deliberately glossed over; and on whether the statistic really does contribute to settling the argument, *or only seems to do so*. The possibility that the presenter's stated motive for quoting the statistic may not be the real motive is a further factor to keep in view.

Questions

Question 5.1 (A)

'Does she have a temperature?' is the colloquial inquiry about someone who complains of feeling unwell. The temperatures 37.0°C or 98.6°F are associated with this question. Where do these numbers come from? Are they accurate? Are they meaningful?

Question 5.2 (A)

According to the Wikipedia article titled *List of sovereign states and dependent territories by population density*, online at [5.8], the population density in persons per square kilometre, at the time of writing, is 409 in the Netherlands, 22 in Chile, 24 in Brazil and 3 in Australia. How accurate are these statistics? Is it meaningful to compare them?

Question 5.3 (A)

School reports for primary school children, giving a child either a 'pass' or a 'fail' grade for each subject, together with a lengthy verbal description of the child's capacity to perform subject-specific tasks, were much in the news in Sydney, Australia a few years ago. Such reports were described by some parents as 'impossible to understand'. The writer of a *Letter to the Editor* of a leading Sydney newspaper at that time was supportive of a government proposal to replace such reports by a simple letter grade assessment of performance in each subject, on an A to E scale. He wrote (we have condensed the wording slightly):

> 'The A-to-E ranking is the most logical form for a school report because most people are "average" or C on any measure (as is shown in the bell curve). It is also educationally justified because it does not label below-average students perpetual failures and gives them some hope of improving. But the problem with any method of grading or ranking is that we cannot know to what extent it can be generalised to the whole state or nation. Anyone who believes that A-to-E ranking of their child in a particular school is an indication of their state or national peer group relationship is being naïve. Aspirational parents send their children to independent [i.e. private] schools because they are keen to see them progress at above the average rate to obtain A or at least B.'

What critiques can you make of the statistical thinking explicit or implicit in this letter?

Question 5.4 (A)

Good statistical work requires more than trustworthy data and methods of analysis. The statistician's citation of the work of others must also be trustworthy. In many statistics textbooks, one can find this quotation from a 1903 book by H.G. Wells titled *Mankind in the Making*: 'Statistical thinking will one day be as necessary for efficient citizenship as the ability to read and write'. This is, however, a very inaccurate quotation. What did Wells actually write on this theme?

Question 5.5 (A)

In general, an average alone provides a minimally meaningful picture of a frequency distribution. However, for some shapes of frequency distribution, the average may actually be quite uninformative, because few of the observed values are actually at or near the average. Can you give some real-world examples?

References

Online

[5.1] http://www.rss.org.uk
[5.2] http://iase-web.org/islp/
[5.3] http://cdiac.ornl.gov/trends/co2/
[5.4] http://ds.data.jma.go.jp/gmd/wdcgg/
[5.5] http://www.esrl.noaa.gov/gmd/ccgg/trends/global.html#global
[5.6] http://www.abs.gov.au/ausstats/abs@.nsf/mf/6202.0. Click the Downloads tab and then the pdf button under the heading 'Publications'.
[5.7] http://www.unece.org/stats/links.html#NSO
[5.8] http://en.wikipedia.org/wiki/ List_of_sovereign_states_and_dependent_territories_by_population_ density

6

Let's hear it for the standard deviation!

Every day there are averages mentioned in the press and other popular media. Occasionally, an average is accompanied by some information about the frequency distribution from which it was calculated, but mostly only the average is reported.

Here is a typical example, quoted from a newspaper: 'It takes an average of 17 months and 26 days to get over a divorce, according to a survey released yesterday. That's the time it takes to resolve contentious issues, such as child custody, property problems and money worries.' Even supposing that this survey was done in a way which permits valid generalisation to the whole community, and that we knew how this average was calculated, it is quite obvious that the average alone provides an incomplete and minimally informative picture of the time it takes people, *in general*, to get over a divorce.

To get something more useful from the survey results, we need to know, in addition – at the very least – the number of respondents to the survey, and some measure of the spread of values in the sample of the variable being studied. While the number of respondents is sometimes mentioned in media reports, a measure of the spread of the data on the survey variable is hardly ever given.

Why might this be? And what could be done to remedy the situation?

Let us start by reviewing how beginning students of statistics build their knowledge of the subject, and note some interesting sidelights along the way.

Anyone who has studied statistics at senior school level or beyond knows that a useful way of summarising a large set of quantitative data is by grouping the individual values in a frequency distribution. It is then a statistician's

A Panorama of Statistics: Perspectives, Puzzles and Paradoxes in Statistics, First Edition.
Eric Sowey and Peter Petocz.
© 2017 John Wiley & Sons, Ltd. Published 2017 by John Wiley & Sons, Ltd.
Companion website: www.wiley.com/go/sowey/apanoramaofstatistics

basic concern to find ways of representing the data in the frequency distribution in a way that is more immediately informative than the detailed frequency distribution table itself. One such representation is a graph of the frequencies, generally in the form of a histogram or frequency polygon. From this graph it is easy to get an impression of three fundamental characteristics of the frequency distribution: where its centre is located, how spread out the data values are relative to the centre, and whether or not the values are distributed symmetrically about the central value.

Each of these characteristics is quantifiable, and the measures used are termed measures of *centrality* (or of *central tendency* or of *location*), measures of *dispersion* (or of *spread* or of *scale*), and measures of *skewness* (or of *asymmetry*), respectively. (There are other quantifiable characteristics of a distribution, too, but their measures are less practically significant than the three just stated.)

If we want to compare two frequency distributions where the data have the same units of measurement, it is most obviously useful to compare them in their entirety (for example, by overlaying one graph with the other, ensuring that the scales on the horizontal axes are the same). But if neither the distribution tables nor their graphs are available, the best remaining option is to compare the distributions on their values of the three descriptive measures just mentioned.

Students learn that there are alternative measures of centrality, of dispersion and of skewness. Each such measure has strengths and weaknesses; no measure is universally 'best'. What is appropriate depends on the context in which the statistical work is being done, and on the level of statistical literacy of the audience being addressed. Two measures of centrality are common in textbooks: the arithmetic mean and the median (the mode is often included as well, but it is not strictly a measure of centrality, as the example '17 is the mode of the data set 7, 8, 11, 14, 17, 17, 17' makes clear). Four measures of dispersion are mentioned routinely: the range; the semi-interquartile range; the mean absolute deviation; and the standard deviation. For skewness, there are two common measures: the quartile measure and the moment measure.

Beginners in statistics generally have little trouble understanding the arithmetic mean, since they have known about it under the name 'average' from childhood. After all, *everyone* knows what an average is! However, not everyone knows that, in the technical vocabulary of statisticians, the term 'average' doesn't refer only to the arithmetic mean. Rather, it is used as yet another generic name for 'a measure of centrality'. It is also little known that all averages that are called 'means' form a family with this unifying characteristic: they each involve *combining*, in some arithmetic way (e.g. by addition or multiplication), the specific values that are being averaged. A rich diagrammatic perspective over several members of the family of means – including the arithmetic, geometric and harmonic means – is found in Lann and Falk (2005).

The median, by contrast, is a 'positional' measure – that is, it occupies the middle position after all the numerical values have been *ordered* from smallest to largest (but in no way *combined*). Its straightforward derivation means that it, too, is easily understood by beginners.

What about measures of dispersion? Researchers in statistics education consistently find that the standard deviation – the measure most commonly used in professional work – is not easily understood by beginning students. It is a difficult concept for at least two reasons.

Firstly, the way the standard deviation is defined seems complicated, and doesn't accord with intuitive notions of variability, as is well illustrated in Loosen *et al.* (1985) and Pingel (1993). Secondly, it seems quite abstract – while it is easy to point to the mean on the graph of a frequency distribution, one cannot point to the standard deviation.

To overcome the first of these obstacles to understanding, it is essential that beginners understand why deviations from the mean, in the formula for the standard deviation, are squared, and why the square root is taken of the mean squared deviation. It is important also to be clear that the standard deviation measures the dispersion of a set of values *around a central value – the arithmetic mean.* An alternative explanatory path – presenting the standard deviation as related to a measure of variability among the values themselves, *without reference to a central value* – is more roundabout, though some people find it more intuitive. See QUESTION 6.1.

Giving the standard deviation practical meaning can help resolve the second difficulty. One way to do this is to rescale the horizontal axis of the graph of a frequency distribution into 'standard deviation units', and then to call on Chebyshev's Inequality. When students understand that, quite generally, no more than one-quarter of the values in a frequency distribution lie beyond two standard deviations from the mean, no more than one-ninth of the values lie beyond three standard deviations from the mean, and so on, the standard deviation becomes something 'tangible'.

When it comes to interpreting and generalising survey information based on random sampling, students learn that, on various criteria, the sample mean produces the 'best' point estimate of the corresponding population mean, but that no reliable judgment can be made about how accurate that point estimate is likely to be without knowing the size of the standard error of the sample mean. It is relevant to mention this here, because the standard error of the sample mean is nothing but a standard deviation – the standard deviation of the sampling distribution of the sample mean. Alas, the concept of a sampling distribution is yet another notion that beginners in statistics find difficult to comprehend.

It's clear now why the standard deviation is so rarely encountered in the popular media: statistical literacy in the community is too low to make it constructive information. Our experience is that this is as true, generally, of journalists as it is of their audiences.

So, let's hear it at last for the standard deviation! Or the semi-interquartile range. Or, indeed, *any* measure of dispersion.

To improve matters globally, it is a good start to act locally. We propose that you, our readers, cease passively accepting media reports that quote the average of a data set – often even without clarifying *which* average – while giving no idea of the associated standard deviation (or standard error, as appropriate).

Engage with journalists and their editors. Emphasise to them how utterly minimal is the useful information about an entire distribution to be found in an average alone. Point them to the particularly striking illustration in Stephen Jay Gould's 1985 essay, online at [6.1]. (Gould highlights his initial despair, as a statistical layman, when his doctor quoted only the median survival time after diagnosing him with cancer.) There is another telling example in QUESTION 1.4.

Encourage journalists to go back to the source of the statistics they are reporting and to present a more informative picture, educating the public as they go. Draw their attention to tutorial resources in basic statistics and statistical literacy, both in print and online. Here are some good examples: Best (2001); Blastland and Dilnot (2007); Royal Statistical Society, Resources for Journalists, online at [6.2]; Robert Niles' website, 'Statistics Every Writer Should Know', at [6.3]); (US) National Council on Public Polls, '20 Questions A Journalist Should Ask About Poll Results', online at [6.4]; and open-ended assistance for journalists via the American Statistical Association, online at [6.5].

Everyone knowledgeable in statistics has a part to play in raising the statistical literacy of our community.

Questions

Question 6.1 (A)

The textbook defines the standard deviation, a measure of the dispersion of the data in a population of N observations X_i ($i = 1, 2 \ldots N$) with mean μ, as $\sqrt{[\Sigma(X_i - \mu)^2/N]}$.

'Why should dispersion be measured relative to a central value?', you may wonder, 'It's the variability among the observations themselves that we should be trying to capture.' Let's consider this idea for a population of three observations, X_1, X_2, X_3. Then an intuitively meaningful measure of variability is $\sqrt{[\sum_i \sum_j (X_i - X_j)^2 / 3]}$, where the summation is over all three pairings of $i = 1, 2, 3$ and $j = 1, 2, 3$ with $i < j$.

For the three observations, evaluate algebraically $A = \Sigma(X_i - \mu)^2/3$ and $B = \sum_i \sum_j (X_i - X_j)^2 / 3$. What is the relation between A and B?

Question 6.2 (A)

For a population of N values, X_i ($i = 1, 2 \dots N$) with mean μ, the variance is defined by $\Sigma(X_i - \mu)^2/N$, that is, the square of the standard deviation. Does deletion of the (single) largest value in any set of N values always reduce the variance?

Question 6.3 (A)

A BBC report of 15 August 2009, online at [6.6], brings together survey data on the average floor area in square metres of new homes (i.e. houses and flats) recently built in seven countries. The areas are UK 76, Ireland 88, Spain 97, France 113, Denmark 137, Australia 206, USA 214. No standard errors are reported. What interesting questions could be answered had the standard errors been quoted? What further information would you like to have from the surveys in order to answer some even more interesting questions?

Question 6.4 (B)

Statistics students soon learn that the sampling variance of the mean of n values sampled with replacement from a normal distribution with mean μ and variance σ^2 is σ^2/n. But what is the corresponding result for the sampling variance of the median?

Question 6.5 (C)

Beginners in statistics are often bewildered by the textbook definition of the 'sample variance', $s^2 = \Sigma(X_i - \bar{X})^2/(n - 1)$, because it is so counterintuitive. (*Note*: n is the sample size and $\bar{X}$ is the sample mean.) This definition is often accompanied by an assurance that 'the reason for division by $(n - 1)$ will become clear when we come to the topic of estimation'. It leaves many learners puzzled. Their intuition is to define the sample variance as the sample analogue of the population variance, $\hat{\sigma}^2 = \Sigma(X_i - \bar{X})^2/n$.

Let's see what we can discover 'when we come to the topic of estimation.' If the X values are sampled randomly, with replacement from a population with mean μ and variance σ^2, the standard error of $\bar{X}$ is given by $\sigma/\sqrt{n}$ and, therefore, the variance of $\bar{X}$ is σ^2/n. If σ^2 is unknown, it could be estimated by s^2 or by $\hat{\sigma}^2$. It can be proved that $E(s^2) = \sigma^2$ and, if the population is normally distributed, that var $(s^2) = 2\sigma^4/(n - 1)$. Also, $E(\hat{\sigma}^2) = [(n - 1)/n]\sigma^2$ and, if the population is normal, var($\hat{\sigma}^2$) $= 2\sigma^4(n - 1)/n^2$.

(Proofs of the results relating to s^2 and $\hat{\sigma}^2$ can be found, for example, in Wilks (1962), pages 199–200, and Cramer (1958), pages 347–349, respectively.)

What do these results say about the relative optimality of s^2 and $\hat{\sigma}^2$ for estimating σ^2 on the criteria of (i) unbiasedness and (ii) efficiency? What conclusion do you draw about which is the 'better' estimator of the variance of $\bar{X}$ in small samples drawn from a normal distribution?

References

Print

Best, J. (2001). *Damned Lies and Statistics: Untangling numbers from the media, politicians and activists.* University of California Press.

Blastland, M. and Dilnot, A. (2007). *The Tiger That Isn't: Seeing through a world of numbers.* Profile Books. (US edition: *The Numbers Game: The commonsense guide to understanding numbers in the news, in politics and in life.* Gotham Books, 2008).

Cramer, H. (1958). *Mathematical Methods of Statistics.* Princeton University Press.

Lann, A. and Falk, R. (2005). A closer look at a relatively neglected mean. *Teaching Statistics* **27**, 76–80.

Loosen, F., Lioen, M. and Lacante, M. (1985). The standard deviation: some drawbacks of an intuitive approach. *Teaching Statistics* **7**, 2–5.

Pingel, L.A. (1993). Variability: does the standard deviation always measure it adequately? *Teaching Statistics* **15**, 70–71.

Wilks, S.S. (1962). *Mathematical Statistics.* Wiley.

Online

[6.1] Gould, S.J. (1985). The median isn't the message. *Discover* **6**, 40–42. At http://www.cancerguide.org/median_not_msg.html

[6.2] http://www.statslife.org.uk/resources/for-journalists/

[6.3] http://www.robertniles.com/stats/

[6.4] http://www.ncpp.org/?q=node/4

[6.5] http://www.stats.org

[6.6] http://news.bbc.co.uk/2/hi/uk_news/magazine/8201900.stm

7

Index numbers – time travel for averages

'Everything seems to be getting dearer. I don't know how I'm going to be able to make ends meet!' This customer complaint has been heard for years in supermarkets and shopping centres. Well, how much dearer *is* everything getting? To answer this, you need to know how statisticians measure the general change in prices over time.

There are many alternative ways of defining such a measure; there is no single 'correct' way. Choosing the most informative measure in any particular setting needs an understanding of the strengths and weaknesses of each of the alternatives. This Overview shows how the statistician's choices mushroom out of what may seem, at first, to be a quite uncomplicated problem.

As the options multiply, it's good to know that the measures are all never far from that most basic of tools for summarising a set of data – the average. They are all, indeed, averages travelling through time.

Yet, unexpected complications arise when defining these measures. For instance, averaging *proportional changes in the values* of a variable between two points in time should, you might think, be done in the same way as averaging *a set of values* of that variable at one point in time. It might surprise you, then, that the average that is most appropriate for the former purpose is often not the one you might think of first ... or second!

To explore these matters further, let us specify the context a little more precisely. Since it is retail customers who are grumbling, we shall assume that we are talking about consumer goods – both durable goods (such as cars, TVs, laptop computers and kitchen appliances) and non-durable goods (such as food, clothing, magazines and soap) – and services (such as public transport, medical consultations, electricity supply and home insurance). From here on, we shall use the term 'products' to include both goods and services.

A Panorama of Statistics: Perspectives, Puzzles and Paradoxes in Statistics, First Edition.
Eric Sowey and Peter Petocz.
© 2017 John Wiley & Sons, Ltd. Published 2017 by John Wiley & Sons, Ltd.
Companion website: www.wiley.com/go/sowey/apanoramaofstatistics

How shall we begin? Given that actual retail price paid is the obvious measure of costliness to the customer, and that retailers compete for customers, we must look at the price of each product, not just in one particular shop but in many shops. We must take care each time to price the same quantity and quality of a product and, preferably, also the same brand (for manufactures), or species (for raw foods), or type (for services) of that product. We must also ensure that any price discount offered is recorded. To condense the volume of data after lots of prices for the same product in different shops have been collected, it will be convenient to use an average price as a representative cost. The arithmetic mean will serve well for this average.

How can we tell whether or not a product is getting dearer over time? Clearly, it will be getting dearer if the ratio of its (average) price in a subsequent period (technically termed the 'current period') to its (average) price in an initial period (technically termed the 'base period') is greater than one. (We use the ratio, rather than the difference, of prices because we are aiming at a measure that has no units; the difference of prices is measured in units of currency.)

If we were to construct such price ratios for many different products, using the same base period and current period for each, what should we expect to find? Experience tells us that some products will have become dearer over time and some cheaper. For example, cars and home insurance will very likely be dearer, while laptop computers and microwave ovens will be cheaper.

There is something else we will be aware of: the quality of most products will have risen over time, and that will be true even of many products that have become cheaper. What explains this steady improvement in observed quality? Partly, it is due to steady advances in science and technology, and in industrial and commercial design; and partly, it is due to market competition, which spurs manufacturers to innovate with product features that outdo what their rivals can offer. How can it be that products of higher quality become cheaper? This could follow from expanding local sales to national and even international markets, thus gaining the benefits of economies of scale on the unit cost of production. It could also be the result of shifting production to a country where the pay structure is lower.

What is the next step in measuring the extent of price changes through time in retail products generally? There are subtleties here that can best be brought out by using some symbols.

Suppose we limit ourselves to considering n products. Denote the price of the ith product in the base period by p_{0i} and in the current period by p_{1i}. The values of i run from 1 to n. We will assume, for the moment, that the product's quality does not change between the two periods. Keep in mind that what we are here calling *the* price of the ith product is, as already explained, actually the average of prices of that product recorded in a number of shops. The price ratio is p_{1i}/p_{0i}. For different products, this ratio may be greater than, equal to, or less than 1.

Then, a natural way to represent the general drift of prices between periods 0 and 1 is to average the price ratios over all the n products. For this average, we prefer a mean to the median, because the median is insensitive to outliers in the set of price ratios – while it is precisely the outliers (relating to the products with the biggest price changes) that are of the greatest concern to consumers.

But which mean? Non-statisticians are likely to say the arithmetic mean (AM) of the ratios: $(1/n)\left[\sum_1^n \left(p_{1i}/p_{0i}\right)\right]$, because that is the one they are most familiar with. But is that actually the best choice? Perhaps unexpectedly, the answer is no. A simple example will show why.

Suppose that, over a given length of time, the price of a loaf of bread doubles, from \$2 to \$4, while the price of a kilogram of butter halves, from \$6 to \$3. Then, in the base period (before there were any price changes), the notional AM of price ratios was $(1/2)\,[(2/2) + (6/6)] = 1.00$. However, after the price changes, the AM of price ratios is $(1/2)\,[(4/2) + (3/6)] = 1.25$. So, for these two goods, the AM shows a logically unjustifiable (and hence spurious) 25% increase in the general price level. The increase is intuitively unjustifiable, as well, for intuition insists that, 'if the price of one good doubles and the price of the other halves, then surely the general price level is unchanged'.

Fitting better with intuition is the geometric mean (GM) of the price ratios. To write the GM of price ratios in symbols, we use the upper-case Greek letter pi (Π), which corresponds to the first letter of the English word 'product'. The expression $\Pi(\dots)$ means 'find the product of the terms in the brackets'. This parallels the sigma notation $\Sigma(\dots)$, which means 'find the sum of the terms in the brackets' (the Greek sigma corresponds to the first letter of the English word 'sum'). The GM is $\left[\prod_1^n (p_{1i}/p_{0i})\right]^{\frac{1}{n}}$. After the price changes, our numerical example shows $\sqrt{\left(\frac{4}{2}\right)\left(\frac{3}{6}\right)} = 1.00$, signifying a stable general price level, as intuition dictates. From this, we understand that the GM is more appropriate than the AM for averaging ratios.

A measure of the general change in prices, which is a pure number (i.e. it has no units), is called a *price index*. The AM and the GM formulae above are examples of simple price indexes (also called 'indices').

So far we have defined a 'general change in prices' index in terms of *an average of price ratios*. We first form the price ratio for each of n products, and then average these n ratios.

---oOo---

Let's look next at a logical alternative to this approach to measurement. Consider closely the following subtle contrast in wording. The alternative approach is a 'change in general prices' index, defined as *a ratio of price averages.*

For a 'change in general prices' index, we first average the n prices in period 0, then average the n prices in period 1, then form the ratio of these two average prices. If we use the AM, this alternative approach produces the price index: $\dfrac{\sum p_{1i}}{n} / \dfrac{\sum p_{0i}}{n}$, which can be simplified to $\sum p_{1i} / \sum p_{0i}$. There is, of course, a GM parallel.

The choice that is made between these two alternative designs for a price index – an average of price ratios and a ratio of price averages – can have far-reaching consequences in practice, for they do not necessarily generate the same measured outcomes. See QUESTION 7.1 for some insight on this.

We must now ask whether *any* of these simple price indexes, in fact, provides a satisfactory answer to the question we posed at the beginning of this chapter. We can get a clue about this from the wording of the consumer's spontaneous lament: 'I don't know how I'm going to be able to make ends meet!' The consumer is clearly thinking in terms of his or her household budget – that is, in terms of the range of products the household regularly consumes. The simple price indexes we have explored are not linked to a specific bundle of products that households regularly consume. Thus, they are inadequate for answering our opening question.

---oOo---

To produce a *consumer price index* (CPI), we need first to agree on the composition of the bundle of products that a representative household consumes. The technical term for this bundle is the 'regimen'. Guidance on defining the regimen usually comes from national surveys of household expenditure, covering many different types of household.

Next, we collect price data on the items in the regimen, both in the base period and in the current period. We create a *weighted* price index specifically linked to the regimen, by weighting the price per unit of each product in the regimen by the number of units of that product that the representative household consumes. Since households rarely consume the products in the regimen in the same proportions as time passes, we have to decide whether to weight by the number of units of product consumed in the base period or in the current period. It could produce quite misleading results to use both in constructing a CPI – for example, to weight base-period prices by base-period quantities, and current-period prices by current-period quantities. It is price changes alone that we are seeking to measure. We must avoid mixing up (or 'confounding', in statistical language) changes in price with changes in quantities.

Writing p_{0i} as the base-period price per unit of the ith product in the regimen and q_{0i} as the corresponding base-period number of units of that product in the regimen, and similarly for the current period, one version of a CPI can be defined by $[(\Sigma p_{1i}q_{0i})/\Sigma q_{0i}]/[(\Sigma p_{0i}q_{0i})/\Sigma q_{0i}] = \Sigma p_{1i}q_{0i}/\Sigma p_{0i}q_{0i}$. Summation is over all n products in the regimen. This index formula is in the form of 'a ratio of weighted price averages *using base-period weights*'. The average is the arithmetic mean. It is, in fact, a very widely-used index, and is commonly called the Laspeyres price index, after the German statistician Etienne Laspeyres (1834–1913), who wrote in 1871 to advocate its use.

If, instead, we write $[(\Sigma p_{1i}q_{1i})/\Sigma q_{1i}]/[(\Sigma p_{0i}q_{1i})/\Sigma q_{1i}] = \Sigma p_{1i}q_{1i}/\Sigma p_{0i}q_{1i}$, we have defined a CPI that is 'a ratio of weighted price averages *using current-period weights*'. This index, too, has its keen proponents. It is commonly called the Paasche price index, after the German statistician Hermann Paasche (1851–1925), who pioneered it in 1874. Because it is necessary to empirically re-evaluate the current period quantities for every additional period for which the index is calculated, the Paasche index is more expensive to maintain than the Laspeyres index.

In line with our earlier discussion, one can also define a CPI as 'an average of weighted price ratios'. Here, too, there are two possibilities: one may use either base-period weights or current-period weights. Altogether, then, we have introduced four versions of a CPI – all based on the AM.

By analogy, there are four ways of defining a CPI using the GM: the ratio of weighted price averages and the average of weighted price ratios – in each case, using either base-period or current-period weights. QUESTION 7.3 invites you to discover for yourself the formula for one of these weighted price indexes.

Any one of these eight versions of a CPI is a possible way of measuring the general change in prices. Given that they may yield different values and, occasionally, very different values, what should the statistician do?

---oOo---

This question has preoccupied designers of index formulae for many years. Their response has been (a) to devise some formal tests for a good index formula (with some theorists claiming that the more such tests a formula satisfies, the better it is), and (b) to acknowledge a number of informal selection criteria that have arisen spontaneously.

The formal tests all relate to logical properties of indexes. For example, the 'time reversal' test rests on the intuitively appealing presumption that a good index should have the property that the 'forward' and 'backward' changes it measures over any given time span are *exactly* inversely proportional.

Here is an illustration. Suppose period 0 is the base period, when an index, I, has the value 1.00. Now, looking forward to period 1: if there is a 25% increase between periods 0 and 1 in what I measures, $I_{0,1} = 1.25$. Next, redefine period 1 as the base period, and look back to period 0. Then, the value of $I_{1,0}$ is the solution of this mathematical problem in proportions: $1.25 : 1.00 :: 1.00 : I_{1,0}$. The result is $I_{1,0} = 0.80$. This confirms that $I_{1,0} = 1/I_{0,1}$. Not all index formulae satisfy the time reversal test – for example, neither the Laspeyres nor the Paasche index formula does. Since there are several formal tests, and no index that satisfies them all, there always seem to be index formula theorists arguing that 'we must keep looking'.

Some informal selection criteria relate to the practicalities of acquiring, in the time available, all the trustworthy data needed for a particular index formula. (This is, incidentally, one reason why the official Australian CPI is published quarterly, rather than monthly.) Other informal selection criteria seek to ensure that the favoured index generates values with a meaningful interpretation. We have already seen, for instance, how using the AM to construct a simple price index generates spurious results. This can partly explain a widespread preference among applied statisticians for CPI formulae that involve the GM, rather than the AM.

Here is a further problem that bedevils the use of either the Laspeyres or the Paasche price index in practice. Over a time span in which the general price level is steadily rising, the Laspeyres price index tends to overstate the true average price rise, and the Paasche price index tends to understate it. Why? Because when prices rise across a large range of products, consumers often protect the purchasing power of their incomes by switching their purchases away from products that are in the (base period) regimen of the Laspeyres price index. Thus, the impact of lower consumption, over time, of items in the Laspeyres regimen means that the general price movements generated by the Laspeyres formula no longer reflect the price movements that consumers are actually experiencing. Analogous, but converse, reasoning applies in the case of the Paasche index.

This phenomenon makes it appealing (at least, in theory!) to consider some kind of compromise between the Laspeyres and Paasche indexes, in order to arrive at index values closer to the true general price rise. One such compromise is the AM or the GM of the two indexes – more usefully the latter, for the reason we have already given. Another approach is represented by the Marshall-Edgeworth price index (see QUESTION 7.5). Needless to say, the substantially increased data collection effort (and cost), when both indexes are to be evaluated, works against the theoretical ideal.

These various issues by no means exhaust the complexity of deciding on an informative measure of the general change in consumer prices. We have not investigated here many elaborate variants of the foregoing CPI formulae. We have also given no attention to the alternative ways there are for periodically updating the regimen of a price index – to allow both for quality change in the included products, and for the eventual introduction to sale or withdrawal from sale of specific branded items.

If you are interested in finding out more about these and other aspects that lie outside the scope of this Overview, you are likely to find a discussion of them among the publications of your national statistical authority. You can locate these authorities online, worldwide, by clicking on the links at [7.1]. For us in Australia, the authority is the Australian Bureau of Statistics, and we can recommend its publication no. 6461.0, titled *The Consumer Price Index: Concepts, Sources and Methods.* The latest edition can be read online on the ABS website at [7.2], or downloaded there without charge. A well-written and not highly technical book on index numbers in general is Crowe (1965).

Questions

Question 7.1 (B)

For the numerical data on bread and butter prices given in this chapter's Overview, evaluate the simple AM and GM price indexes in the form of a ratio of price averages. Comparing the indexes calculated using the AM – (i) average of price ratios, and (ii) ratio of price averages – do you have reasons to prefer one of these to the other? What can you say about the corresponding GM indexes?

Question 7.2 (B)

In the context of QUESTION 7.1, we wrote in the Overview '[f]itting better with intuition is the geometric mean (GM) of the price ratios.' The demands of intuition appear to be determining the statistician's choice of best simple price index. Isn't that an unusually powerful role for intuition in statistical theory?

Question 7.3 (B)

What does the formula for a *ratio of weighted price averages* look like when the weights are base period quantities and the average is the geometric mean? (This is the GM analogue of the AM-constructed Laspeyres price index.)

Question 7.4 (B)

Financial advisers traditionally enthuse about the 'wonders' of compound interest. 'Imagine', they will say, 'that you have invested $100 every month over 40 years at 10% per annum, compounding monthly. At the end of this period, you will have put in $48,000 of your own money. But with the wonders of compound interest, these $48,000 will have grown to a staggering $632,408. That means your own capital has been multiplied about 13 times.'

This calculation is numerically correct, but unfortunately the statistic in the previous sentence is wildly misleading in practical economic terms. Why?

Question 7.5 (B)

The Marshall-Edgeworth CPI index uses, as its weights, the AM of the base period and current period quantity weights that feature in the Laspeyres and Paasche indexes, respectively. The formula for this index is $\Sigma p_{1i}(q_{0i} + q_{1i})/ \Sigma p_{0i}(q_{0i} + q_{1i})$, where the summations extend over all n items in the regimen. The index is named in recognition of two of its advocates, the English economist, Alfred Marshall (1842–1924), and the Anglo-Irish economist and statistician, Francis Ysidro Edgeworth (1845–1926). You will find further mention of Edgeworth in FIGURE 22.1.

In this chapter's Overview, we wrote: '... we have to decide whether to weight by the number of units of product consumed in the base period or in the current period. It could produce quite misleading results to use both in constructing a CPI – for example, to weight base-period prices by base-period quantities and current-period prices by current-period quantities. It is price changes alone that we are seeking to measure. We must avoid mixing up (or 'confounding', in statistical language) changes in price with changes in quantities.'

The Marshall-Edgeworth index clearly weights p_0 in part by q_0 and p_1 in part by q_1. Does that mean that this index confounds price change and quantity change?

References

Print

Crowe, W.R. (1965). *Index Numbers – Theory and Applications.* Macdonald and Evans.

Online

[7.1] http://www.unece.org/stats/links.html#NSO

[7.2] http://www.abs.gov.au/ausstats/abs@.nsf/mf/6461.0. Click the Downloads tab and then the pdf button under the heading 'Publications'.

8

The beguiling ways of bad statistics I

By bad statistics, we mean the incorrect or inappropriate use – either from insufficient knowledge or from a calculated intention to mislead – of statistical data (and data displays) or statistical methods, in support of some arguable proposition, such as a scientific hypothesis, an advertising claim or a political point of view.

Many people find statistical data fascinating – even bewitching – when they understand the context. The long-time popularity of *Guinness World Records* and *Wisden Cricketers' Almanack* is clear evidence of this. However, even when they do not fully understand the context, the general public are still receptive when statistics are presented. True, that receptiveness is sometimes too trusting but, increasingly, people are resisting the notion – perhaps (mis)remembered from school arithmetic lessons – that 'there's no arguing with a number, especially a very precisely stated number.'

That is the bright side for everyone who would like to see published statistics better engaged with in our society.

But there is a darker side. People who know only a little about the principles of statistics may unintentionally mislead the public with inappropriate statistics or erroneous statistical arguments. More troublingly, there are individuals with a vested interest in capturing people's trust who knowingly misuse statistics to bamboozle others for questionable ends.

How convenient it is that English has a single word that bridges the bright side and the dark side of statistics, with their very contrasting qualities – bewitching and bamboozling. That word is *beguiling*. Thus, our theme in this chapter and the next is *the beguiling ways of bad statistics*.

---oOo---

A Panorama of Statistics: Perspectives, Puzzles and Paradoxes in Statistics, First Edition.
Eric Sowey and Peter Petocz.
© 2017 John Wiley & Sons, Ltd. Published 2017 by John Wiley & Sons, Ltd.
Companion website: www.wiley.com/go/sowey/apanoramaofstatistics

We wrote in CHAPTER 1 about the value of statistical methods, appropriately used, for finding reliable meaning in data collected from the uncertain world that is our constant reality.

It is quite another matter if statistical data or statistical methods are used inappropriately to underpin an argument. At the least, such inappropriate use may simply yield obvious nonsense. An amusing example is the complaint made by a defendant on a drink-driving charge. 'I've looked at the data,' he said, 'and I've seen that 95% of road accidents are caused by sober drivers. Those sober drivers should really get off the road and let us drunk drivers get around in safety.' (You'll find further examples in QUESTIONS 1.3 and 9.1.)

More concerning are situations where the inappropriate use of statistics produces plausible but incorrect conclusions. Such a situation may arise from ignorance of what is appropriate. Sometimes, however, deliberate misuses of statistics must be suspected. As you should expect, such misuses are not the work of ethical professional statisticians. Rather, they are the machinations of a regrettable minority among skilled persuaders – people whose goal it is to bring the uncommitted around to a particular point of view, and who are not too concerned about using a little deception to achieve that goal. You will probably find it easy to bring to mind examples of deceptive statistics you have seen or heard about. It is such examples that get the public saying, cynically, 'you can prove anything with statistics!'

In CHAPTER 5, you will find our recommended approach to detecting deceptive statistics. It rests on adopting a sceptical frame of mind and asking four questions. Who produced these statistics? How were they produced? Who is presenting them to me? And why?

Probing in this way, you can become aware whether self-serving interests lie behind the way the data were collected or summarised. You may discover that there has been fudging of the figures, or of the concepts behind the figures. You may also reflect on whether there are perhaps other interpretations of the statistics than those that are being urged on you.

---oOo---

Let's look now at some inappropriately used statistical methods. They are all 'tried and true' techniques for producing misleading conclusions, so a little irony from our side doesn't seem out of place!

It's important to impress, and a big statistic is always more impressive than a small one

An easy way to generate a big statistic is to quote the maximum of a set of values, rather than some typical or fairly representative value. This is a favoured approach in the news media. We may learn, for example, that 'up to 60 people were killed on each occasion in train crashes around the world last year'. That sounds absolutely horrific. Perhaps no more than 5 people were killed on each occasion, except for one occasion when 60 people were killed. The report is not, strictly speaking, inaccurate, but the interpretation we are likely to give it is certainly misleading.

To discuss a frequency distribution, all you need to do is talk about the average

We emphasise, in CHAPTER 6, the inadequacy of an average as the sole descriptor of a frequency distribution. In the absence of a drawn histogram, one can still get a fair idea of a distribution's position and shape from a set of three measures: its average, its dispersion about the average, and its degree of skewness (i.e. departure from symmetry). Why are measures of dispersion and skewness so routinely missing in popular journalism? Because the community's level of statistical literacy is not generally high enough to make these two measures informative – and, sadly, this is true everywhere. QUESTIONS 1.4, 5.5 and 6.3 show how easily misleading it is to provide only an average, when the context makes it clear that a fuller description of an entire distribution is important. Needless to say, this state of affairs suits very well anyone who has a motive for withholding the fuller picture.

League tables are a great way to show who's simply the best … and who's not

An institutional league table is a ranking of a set of (usually competing) institutions (e.g. schools in a district, corporations in an industry), according to some (usually quantitative) criterion of excellence. If the table is 'simplified' by suppressing each institution's criterion score (as is often the case in popular reporting), all that is left is a list that reads simply: 1. A, 2. B, 3. C and so on, where A, B, C … are names of institutions. Superficially, it looks as if A is 'simply the best', and one really wouldn't want to have anything to do with V, N, or even H, lower down the table! However, just as discussing a frequency distribution by referring only to the mean ignores valuable information about the spread of the data, so publishing a 'simplified' league table conceals the valuable information to be found in the sizes of differences between criterion scores. If the criterion scores of institutions A and B (for example) are very similar, those institutions are, to all intents and purposes, *both* 'simply the best'. Further revelations on the limitations of league tables can be found in Goldstein and Spiegelhalter (1996).

League tables (especially 'simplified' ones) are beloved by institutional propagandists who want to promote their own institution over others. They will, of course, seek to influence the choice of ranking criterion, so that their institution will come out as high as possible in the ranking. All the more reason, then, for outside observers to ask themselves 'is the criterion being used to rank the institutions the most appropriate one *from my point of view*? If not, how different would the ranking look if a more appropriate criterion were used?' See QUESTION 3.2 for an illustration.

Generalisations from a survey are reliable, irrespective of the response rate and of whether the sample is random

This is the subliminal message of countless survey and poll reports, published in all the popular media, to justify the claims of some advocacy group or other. Statistically informed observers will recognise the failure to report the response rate or the sampling scheme as the deceptive stratagem that it often is. No statistical method can be correctly used to make reliable inferences about a population from the information in a non-random sample, such as a self-selection sample, a quota sample, or a convenience sample (the answer to QUESTION 3.3 has more detail on this point). Similarly, a high rate of non-response – especially if the cause of non-response might be related to the objective of the survey – makes any generalisation unreliable.

---oOo---

These are four among the most commonly met ways in which disingenuous persuaders misuse statistics. There are more in CHAPTER 9, and even more in several books whose authors have concentrated specifically on highlighting the damage that may be done by bad statistics.

The most widely known of these books is also one of the earliest – the whimsically titled *How to Lie with Statistics* by Darrell Huff (1954). This lively book, reprinted multiple times, and now among the 'classics' of the statistics literature, presents many memorable real-world examples of inappropriately used statistical methods and deceptive statistics. More recent books in the same spirit, but revealing possibilities for deception on a far wider scale, are Spirer, Spirer and Jaffe (1998), Best (2001) and Best (2008). Further current examples can be found in the blog by David Spiegelhalter and colleagues at the Understanding Uncertainty website, online at [8.1].

Questions

Question 8.1 (A)

It seems that people have been fascinated by numerical statistics for a long time: *Wisden Cricketers' Almanack*, for example, began publication in 1864. The public's appetite for such numerical facts was, no doubt, also fed by other popular compilations in that era, and perhaps even earlier. But we know of only one contributor who supplied statistics in verse! This 19th century British poet wove precise statistics into many of his poems, which he wrote with a masterly disregard for accurate scansion. Here is a sample of his work (the first, fifth and sixth of 23 stanzas). Who is the poet, and what is the poem's title?

> 'Twas on the 18th of August in the year of 1798,
> That Nelson saw with inexpressible delight
> The City of Alexandria crowded with the ships of France,
> So he ordered all sail to be set, and immediately advance.
>
> …
>
> The French force consisted of thirteen ships of the line,
> As fine as ever sailed on the salt sea brine;
> Besides four Frigates carrying 1,196 guns in all,
> Also 11,230 men as good as ever fired a cannon ball.
>
> The number of the English ships were thirteen in all,
> And carrying 1012 guns, including great and small;
> And the number of men were 8,068,
> All jolly British tars and eager for to fight.

Question 8.2 (A)

Chapter 8 of *How to Lie with Statistics*, by Darrell Huff, is titled 'Post hoc rides again'. 'Post hoc' is part of a longer Latin expression describing a logical fallacy. Explain this logical fallacy. Can you find some real-world statistical examples of this fallacy?

Question 8.3 (A)

A news report says 'Obesity is costing the country $56 billion a year'. Quite apart from the suspect accuracy of the specific number, there is a lack of clarity in the meaning of the statement as a whole. What, exactly, do you understand the statement to mean? How might someone else interpret it differently? Reflecting on these alternative interpretations, what do you conclude about statements of this kind, whatever major medical or social ill they may refer to?

Question 8.4 (B)

The way a questionnaire is constructed – that is, the wording of the questions and the design of their flow – gives scope for manipulating the conclusions that will be obtained. A classic British TV comedy series from the 1980s demonstrated (satirically) how to achieve any desired result in an opinion poll. What is the title of this TV series, what was the political question under consideration, and what questionnaire technique was used to obtain the desired result?

Question 8.5 (A)

On a summer day, a Sydney newspaper reported: 'Food poisoning cases more than double over the summer months as people go back and forth to their fridge and overload it with food, increasing its temperature. A survey … found that some household fridges were twice as warm as they should be after groceries were transferred into them and they took four hours to return to a safe temperature. Harmful bacteria multiply when food is kept [above] 5 degrees [Celsius]. Of the 57 fridges checked in the study, almost 23% had an average temperature of more than 5 degrees … The highest average temperature for one fridge was 9.5 degrees.'

The following day there was a *Letter to the Editor*: 'You report that some domestic refrigerators are "twice as warm as they should be" when overloaded. A temperature change from 5 degrees to 10 degrees does not indicate a doubling of heat content – it represents an increase of about 1.8 per cent.'

What type of variable is being discussed in this exchange? Can you explain why the figure 1.8% is correct for the temperatures stated?

References

Print

Best, J. (2001). *Damned Lies and Statistics: Untangling numbers from the media, politicians and activists*. University of California Press.

Best, J. (2008). *Stat-Spotting: A field guide to identifying dubious data*. University of California Press.

Goldstein, H. and Spiegelhalter, D.J. (1996). League tables and their limitations: statistical issues in comparisons of institutional performance. *Journal of the Royal Statistical Society, Series A* **159**, 385–443.

Huff, D. (1954). *How to Lie with Statistics*. Norton.

Spirer, H.F., Spirer, L. and Jaffe, A.J. (1998). *Misused Statistics*. Dekker.

Online

[8.1] http://understandinguncertainty.org/

9

The beguiling ways of bad statistics II

In the previous chapter, we described four settings in which bad statistics can beguile us. Here, now, are four more.

When assembling evidence to support some controversial proposition (e.g. during a political debate), any number is better than no number

In beginning a public lecture in 1883, the British physicist William Thomson (later Lord Kelvin) offered the following dictum: '... when you can measure what you are speaking about, and express it in numbers, you know something about it; but when you cannot measure it, when you cannot express it in numbers, your knowledge is of a meagre and unsatisfactory kind ...' (online at [9.1]).

Thomson was underlining the indispensability of measurement to the progress of the physical and natural sciences. Today, however, his assertive message is being advanced indiscriminately – bewitching (for example) many of the social sciences and, in particular, the practice of government and public administration.

No social or economic policy proposal is felt to be adequately supported nowadays, unless there are numbers among the evidence. More unfortunately, this sentiment has been reinterpreted by lax thinkers in many areas of public debate to imply that *if there are numbers* among the evidence, *those numbers should suffice to clinch the matter*. From such an illogical position, it is then only a short hop to the view that it is not the intrinsic fitness-for-purpose of the numbers that is the vital thing but, rather, whether the public can be convinced to accept whatever plausible numbers the proponents can find.

A Panorama of Statistics: Perspectives, Puzzles and Paradoxes in Statistics, First Edition.
Eric Sowey and Peter Petocz.
© 2017 John Wiley & Sons, Ltd. Published 2017 by John Wiley & Sons, Ltd.
Companion website: www.wiley.com/go/sowey/apanoramaofstatistics

This situation brings with it two substantial dangers: firstly, that significant factors which cannot be expressed in numbers (but could be perfectly well assessed in qualitative terms) will simply be ignored; and, secondly, that the temptation to bring forward seriously inappropriate numbers or measurements (however plausible they can be made to look) will become irresistible. These twin dangers are neatly encapsulated in a memorable aphorism: not everything that counts can be counted, and not everything that can be counted counts.

Let's look at some examples.

The field of cost-benefit analysis is rich in measurements. After all, the whole idea (in theory) is to measure all the costs and all the benefits of some planned project, and to proceed if the benefits exceed the costs by some pre-specified margin. In order to compare costs and benefits straightforwardly, they need to be measured on a common basis. The technical term for such a basis is a 'numeraire'. The usual numeraire in cost-benefit analysis is money. It will be apparent that (in practice) an elaborate project, such as the installation of a national fibre-optic network to give everyone reliable high-speed internet access, will have many costs and benefits that are simply unforeseeable, and others that are not readily expressible in money terms. For both these reasons, the money values that are ultimately compared may be so inaccurate as to produce a conclusion that could, in hindsight, be seen to have been profoundly wrong.

Similar difficulties arise when economists evaluate the benefit to an individual of a university education. The usual measure of 'benefit' is the difference between two monetary values – the discounted present value of the individual's expected lifetime flow of annual earnings *with* a university qualification, and the parallel calculation for the individual *without* a university qualification. Evidently, this approach ignores entirely a whole array of intellectual, cultural and social benefits that a university education can engender. It might be argued in response that annual income can serve as a proxy for all these non-material benefits – but this is, surely, a flimsy argument.

It should not be surprising that proxies play a prominent role in this era of 'measurement at any price', for unmeasurable concepts abound in every area of public policy making. It is precisely because good measurable proxies are so hard to find that insisting on giving numbers the pivotal role in supportive evidence is so rash.

Several countries have, in recent years, launched initiatives to measure citizens' well-being (or 'happiness', as it is being popularly described). The concept of well-being is easy to comprehend but by no means easy to measure. Proxies are needed. Will they be appropriate?

When a statistic is politically needed but nobody knows its numerical value accurately, it's OK to guess

This phenomenon is observable at every level of perspective. At the international level, we may take as an example the coordinated system of national economic accounts. The Statistics Division of the United Nations has agreed upon a standardised framework of national accounts for all its member nations. However, some countries may be too poor to invest in the elaborate procedures for data gathering that are needed to produce reliable national statistics of any kind. In such cases, those statistics may be wildly approximate. Some consequences are reviewed in Moll (1992). Many more examples, drawn from both developed and developing countries, are found in Alonso and Starr (eds, 1987).

At national level, professional standards within a country's official statistics office tend to constrain the extent of pure guesswork that goes into to the entire range of the statistical series it produces. Nevertheless, 'heroic' assumptions are sometimes made even in the most reputable of these offices. See the UK examples in Giles (2008).

In unofficial contexts, what we might call *suspect statistics* are not hard to find. A generic example, frequently met in the media, is a statistic that summarises the financial burden imposed on the community by the prevalence

'I hear that 73% of statistics are made up on the spot.'

of some disease or social ill. The following figures, recently reported in Australian media, are typical: 'Divorce is costing the Australian economy $14 billion a year', 'Road congestion in Australia's cities could cost the economy $20 billion by 2020' (see also QUESTION 8.3). You can easily find your own society's examples by searching the web for simple variants of the phrase 'costs the economy billions'.

What can be said of such statistics? Given the complexity of the surrounding issues, it is unlikely that they were estimated in some formal statistical way. How much might be due to guessing? We shall leave that to you to reflect upon. Once published, however, such statistics are rarely challenged publicly, and they soon take on a life of their own.

When a persuasive statistic is needed to advance an argument, there is no need to explain on what scale it is measured, even if that scale is unlikely to be familiar to the audience

An ethically principled argument based on (numerical) statistics uses the statistics in a way that is both meaningful to the audience and informative in the context in which it is quoted. To be *meaningful*, a statistic must obviously be appropriate to the purpose of the argument. Additionally, the scale on which the statistic is measured should be familiar to the audience. For example, 'everyone knows' that the body temperature of a healthy person is roughly between 36 °C and 37 °C, and that 41 °C, only four degrees higher, is already dangerously high.

A common way of making a statistic *informative* is to provide a reference base for comparison – for instance, the value of the same statistic in an earlier time period. That explains why percentage changes are more often quoted than absolute values. Even then, there are many possibilities for misunderstanding (see Polito (2014), online at [9.2]).

If the audience is unfamiliar with the measurement scale (or the scale is simply unmentioned), then meaning may be lost. A statistic with little meaning is hardly likely to be informative, even when a reference base is provided.

What is the reality? A scan of newspapers and other publications aimed at the general public readily reveals quoted statistics that are measured on scales very unlikely to be familiar to the public. Yet, in each case, the scale is explained poorly or not at all. Here are two examples we have seen recently.

The degree of inequality of personal income distribution in a country at a point in time can be measured by the Gini coefficient (this statistic is directly related to the Lorenz curve, which appears in QUESTION 23.2). Its values are bounded by zero and one, the former signifying that everyone has the same income, and the latter that all income is in the hands of a single individual. A newspaper report gives the Gini coefficient for Australia in 1994 as 0.336, but explains only that a higher number means a less even income distribution. There is no mention that the Gini scale is bounded by 0 and 1. So when the general reader is told next that the corresponding value for 2013–14 is 0.333, there is no answer to the question in the reader's mind: is that a major change or a minor change?

Global rankings of countries, based on a variety of indexes, are now published annually. Wikipedia has, for instance, entries for the Global Peace Index, the Corruptions Perception Index and the Ease of Doing Business Index, and there are many more (see online at [9.3]). Each of the rankings is determined by the numerical values of the corresponding index. The index values, in turn, are determined by combining scores on a large number of relevant indicators. When any of these rankings is quoted in the media, the underlying index values are rarely given, much less the scale on which the index values lie. To the uninformed reader, the rankings therefore have very little meaning (for reasons explained in more detail in CHAPTER 8). Nor can much meaning be found in the index values (where they are given), since the measurement scale is a mystery to the reader – is the scale bounded; is it linear? In the end, all such rankings and index statistics have simply to be taken on trust. To the general reader, they mean what the journalist says they mean.

Statistical point forecasts are all you need; interval forecasts are too complicated

Media reports are full of forecasts. It is not always clear whether these forecasts were constructed by statistical methods, but it is almost universally the case that they are point forecasts, rather than interval forecasts. Seeing only point forecasts in the media, the public naturally believe that point forecasts are all you need (and, indeed, all that a statistician can possibly provide).

The *precision* of a point forecast is seductive. It bamboozles the mind, distracting it from asking about *accuracy*. A point forecast offers no suggestion that it might be wrong, let alone *how wrong* it might be.

Here is an illustration. Discussing some statistical modelling by the Australian Treasury of the implications of an increase of 14 million in the Australian population by 2050, the editorial of a Sydney daily newspaper quotes government forecasts: 'By 2050, Australia will need an additional 173,000 kilometres of road, 3254 schools, 6.9 million homes, 1370 supermarkets, 685 department stores and 1370 cinemas.' Neither more nor less!

Given that these forecasts rest on statistical methods, an interval forecast would put each of them into perspective. Consider the needed number of schools. For any desired value of the probability that the statement 'By 2050 Australia will need an additional $3254 \pm X$ schools' is true, the statistical interval forecast procedure provides the value of X. Now there is both a prompt that the forecast is uncertain, and an indication of how uncertain it is.

Of course, there are subtleties in both the derivation and the interpretation of an interval forecast (see QUESTION 20.5). However, they are not so subtle as to preclude explanation to the public, if people are functionally literate in statistics.

---oOo---

These are by no means all the beguiling ways of bad statistics. Bad statistical graphics, for example, form a category all of their own. Inappropriate and misleading visual representations of data – especially politically-sensitive data – turn up frequently in institutional reports, journalism and marketing. This is, however, too big a category to explore adequately here. Fortunately, there is a large critical literature. Anyone with a basic knowledge of statistics should find these three books informative: Schmid (1983), Tufte (1983) and Jones (2006). Since 1990, Howard Wainer has written a column titled 'Visual Revelations' in each of issue of *Chance* magazine. Many of these columns critique flawed, and even misleading, statistical graphics that have appeared in public media. Wainer has collected these columns and other essays on related themes in several books, including Wainer (1997) and Wainer (2009).

As CHAPTER 3 makes clear, there is a key to creating resistance to all the negatively beguiling ways of bad statistics. It is advancing the growth of statistical literacy in the community.

Questions

Question 9.1 (A)

A naïve young man at a weekend party drank five shots of whiskey with iced water. The next day he awoke with a hangover. The following weekend at another party he drank five shots of brandy with iced water and had another hangover. At a third party, it was five glasses of white wine with iced water that gave him a hangover. He reflected on this melancholy chain of experiences. 'I'll never drink iced water at a party again', he decided.

What specific cautions does this story imply about the causal interpretation of an observed association between two variables?

Question 9.2 (A)

A television announcer states that there is a 40% chance of rain on the following day. As a person with statistical training, how do you interpret this statement? How might it be interpreted by people without statistical training?

Question 9.3 (A)

Which eminent British scientist published, in 1872, a paper entitled 'Statistical inquiries into the efficacy of prayer'? What was the hypothesis about which the inquiries were made, and what was the conclusion of the investigation?

Question 9.4 (B)

A beguiling statistical strategy commonly used to market an apparently desirable investment (e.g. a house in a well-to-do suburb, or a painting by a famous artist) is to cite the documented purchase price of a closely similar investment at some distant date and its (far greater) recent selling price, and to highlight the percentage return that was achieved. How many ways can you think of in which this information may be misleading to a prospective investor, in judging his or her likely future success with the particular investment being marketed?

Question 9.5 (B)

A commonly used measure of bivariate correlation, r, is due to Karl Pearson. For sample data on two variables X and Y, $r = \Sigma x_i y_i / \sqrt{(\Sigma x_i^2 \Sigma y_i^2)}$, where x_i and y_i are values of X_i and Y_i in deviation from their sample means, and summation is over the number of sample values. The following data are, in order, the height in cm (X), weight in kg (Y) and age in years (Z) of individuals A, B and C: A (180, 86, 45), B (173, 82, 50), C (178, 77, 40). Find $r(X,Y)$ and $r(Y,Z)$.

From these results, what do you conjecture about $r(X,Z)$? Now find $r(X,Z)$. Is this what you expected? What have you discovered?

References

Print

Alonso, W. and Starr, P. (eds, 1987). *The Politics of Numbers*. Russell Sage Foundation.

Giles, C. (2008). Lies, damned lies and befuddlement. *Significance* **5**, 82–83.

Jones, G.E. (2006). *How to Lie with Charts*, 2nd edition. BookSurge Publishing.

Moll, T. (1992). Mickey Mouse numbers and inequality research in developing countries, *Journal of Development Studies* **28**, 689–704.

Schmid, C.F. (1983). *Statistical Graphics – Design Principles and Practices*. Wiley.

Tufte, E.R. (1983). *The Visual Display of Quantitative Information*. Graphics Press.

Wainer, H. (1997). *Visual Revelations*. Erlbaum.

Wainer, H. (2009). *Picturing the Uncertain World*. Princeton University Press.

Online

[9.1] Page 80 in his collected *Popular Lectures and Addresses*, Vol. 1 (1889). At http://www.archive.org/stream/popularlectures10kelvgoog#page/n100/mode/2up

[9.2] Polito, J. (2014). The language of comparisons: communicating about percentages. *Numeracy* **7** (1). At http://scholarcommons.usf.edu/numeracy/vol7/iss1/art6/

[9.3] https://en.wikipedia.org/wiki/List_of_international_rankings and http://www.dataworldwide.org/websites/data_indexes.htm

Part III

Preliminaries to inference

10

Puzzles and paradoxes in probability

As you may already have discovered, there are lots of puzzles and paradoxes in theoretical and applied probability. In this chapter, we want to look a little more deeply into why this subject is so rich in difficulties. First, though, it's good to be clear about who finds work with probability difficult. The answer is ... everybody. Even statisticians! And that has been so for a long time.

In 1654, the French mathematician Blaise Pascal engaged his great contemporary Pierre de Fermat in a joint inquiry on two fundamental matters: how to assign numerical probabilities to chance events, and how to determine the probabilities of compound events. These tasks had always seemed so challenging that no real start had been made on them since, many centuries earlier, a profound realisation had been written into the Bible (Ecclesiastes 9:11): '... time and chance happen to all [mankind]'. However, whereas time has been measured for some 4000 years (these days, with supreme precision and accuracy), efforts to measure chance are hardly 400 years old, *and the concomitant problems are still not solved to everyone's satisfaction.* For a short overview of some of these problems, see Good (1959).

It is mostly the finer aspects of these unsolved problems that give rise to puzzles in probability. Probability theory is full of subtleties. If they are neglected or unrecognised, difficulties of understanding and interpretation soon ensue. There is another obstacle, too – human intuition. Intuition has been neatly defined as 'knowing without much thinking' and (more strikingly) – by Gigerenzer (2008) – as 'the intelligence of the unconscious'. When someone's intuition about something encounters accepted theory, what accepted theory says may come as a shock to his or her intuition. Thus a paradox is born.

A Panorama of Statistics: Perspectives, Puzzles and Paradoxes in Statistics, First Edition.
Eric Sowey and Peter Petocz.
© 2017 John Wiley & Sons, Ltd. Published 2017 by John Wiley & Sons, Ltd.
Companion website: www.wiley.com/go/sowey/apanoramaofstatistics

A paradox is a proposition that (intuitively) 'feels' true (false), though accepted theory shows it to be false (true). To 'explain' (or 'resolve') a paradox is to clarify where, why and to what extent intuition needs to give way to accepted theory. (Another kind of paradox, which we shall not consider further here, arises from a self-contradictory argument before it is recognised to be self-contradictory.) What is, perhaps, unexpected is the stubbornness of human intuition about chance when contradicted by results from probability theory – as our account in this chapter will illustrate.

---oOo---

Let's look now at some concepts in probability theory, and see how they give rise to puzzles and paradoxes – sometimes very perplexing ones, especially to students of statistics.

Many technical terms in probability and statistics are also words used in a somewhat varied sense in ordinary English conversation. This is likely to be a source of confusion to lay people, and the fact that their intuition is rooted in the popular usage of these terms may deepen the confusion. Two such terms are 'random' and 'independent'.

In popular usage, all the following words are ready synonyms for 'random': accidental, arbitrary, chaotic, haphazard, patternless, unordered and unpredictable. The popular notion of randomness is, evidently, a rather fuzzy one!

To a statistician, randomness is a concept that is at once simpler and more complex than the popular notion. The statistician is concerned with randomness in two particular settings:

a) defining randomness (as a theoretical underpinning of the concept of a 'random variable'); and
b) defining operational methods for generating sequences of random numbers (as the basis for selecting a random sample and for many other practical applications).

For these purposes, the required attributes of randomness are '*patternlessness*' and *unpredictability* (other fields, such as physics, mathematics, computer science and psychology, refer to further facets of randomness). Thus, at first sight, the statistician's 'randomness' is simpler and more focused than the fuzzy popular usage. Moreover, we note, achieving patternlessness goes some way towards achieving unpredictability.

But then comes the need to devise practical mechanisms that *it is hoped* will generate patternless sequences of values – and also to define and assess patternlessness in those generated sequences. Otherwise, we cannot judge progress towards our ultimate goal – to produce flawlessly random sequences. The layman's intuition protests at such elaborateness. 'I can *easily* write down a sequence of patternless numbers' is the layman's paradoxical cry. But once such 'amateur' efforts are scrutinised, the layman's claim is generally soon repudiated.

It becomes clear that effective randomness-generating mechanisms need to involve some kind of machine process, running with minimal human input. For the past 80 years, there have been intensive efforts to realise such hardware- or software-based processes. Enormous progress has been made but perfection has not yet been attained.

This activity has, of course, had to proceed side-by-side with technical advances in pattern detection. To give a flavour of the technicality: we need to define the kinds of recurring patterns that we wish to have *absent* from a sequence of random data, then we need to devise reliable tests that will detect each such pattern, if it is present. Since the number of conceivable patterns is vast, you will see that certifying perfect patternlessness is actually an unachievable ideal. An imperfect compromise, in practice, is to test formally only for the presence of 'obvious' patterns.

Even accepting this kind of imperfection (which can, at best, result in 'near-random' sequences), there is yet a further challenge – this time related to unpredictability. Here is the essence of the problem: unpredictability of the next value to be generated (given all the already-generated values) is not assured from patternlessness alone. What is also required is *independence* of the generated values from one another. This is something qualitatively different; whereas patterns are deterministic forms, independence is a probabilistic relation. To take an example: suppose we are trying to generate a long random sequence of the ten digits 0–9. One way of achieving independence is for the generating mechanism to meet the specification that the probability of the next digit generated being (say) a 5 is fixed, regardless of the value of the digit that was generated immediately before.

There are also deeper technicalities in the statistical and philosophical literatures on randomness. They all point to one conclusion: perfectionist pursuit of randomness in data generation can, in practice, fall victim to its very complexity. You will find more about this in CHAPTER 11.

---oOo---

Mention of independence suggests a closer inspection of the concept. Here, too, popular intuition has its own unconscious knowledge: 'things that are independent are unconnected with (*or* unrelated to) each other.' To a statistician, however, this formulation is far too vague to be useful. Indeed, it can also be very confusing. To digress for a moment – many students' confusion of mutually exclusive events with independent events may come from the fact that mutually exclusive events are represented by non-overlapping (and, hence, unconnected) circles in a Venn diagram.

In probability theory, independence of two events, A and B, is defined quite concisely by the relation $P(AB) = P(A).P(B)$, where $P(AB)$, also written $P(A \cap B)$, is the probability of A and B both occurring. In formal terminology, this is 'probabilistic independence' or 'stochastic independence', but it is more commonly called 'statistical independence'. This is to distinguish it from other senses of independence – as are found, for example, in mathematics, logic and popular speech. Unless there is a risk of ambiguity, we shall continue here simply with 'independence'.

The relation $P(AB) = P(A).P(B)$ is both a necessary and sufficient condition for independence. Moreover, it applies even to events whose probability of occurrence is zero. Unfortunately, working with events whose probability is zero creates many puzzles and paradoxes of its own, so it is generally convenient to avoid doing so. For instance, any event with probability zero is independent of itself. (Is that a shock to your intuition?) The proof is quite straightforward. Why not try it before reading on?

If we exclude events with probability zero, the above condition for independence can be transformed into either of the following relations involving conditional probabilities: $P(A|B) = P(A)$ and $P(B|A) = P(B)$. Each of these versions is also a necessary and sufficient condition for the independence of A and B. To students of statistics, these new versions offer something appealing – the opportunity to remake their intuition to 'know' independence the way a statistician understands it. To take the first version, independence means that the probability of event A occurring is the same, *whether or not* event B occurs.

However, *even when they have assimilated this meaning*, many students' intuition remains committed to the popular notion of independence. Thus, if a probability problem involving events A and B has the formal solution that A and B are independent, students may insist that this result is paradoxical if there is any verbal statement within the problem that explicitly *or implicitly* indicates that the events are somehow logically connected.

For example, two coins are tossed randomly, one after the other. Event A is 'the first coin shows a head'. Event B is 'the coins fall alike'. Then, $P(A) = 0.5$,

$P(B) = P(HH) + P(TT) = 0.5$, $P(AB) = P(HH) = 0.25$. Thus $P(AB) = P(A).P(B)$, and so the events are independent. But, intuition dictates, that cannot be! 'When the first coin is a head and we are told the coins fall alike, the second coin no longer has a choice. It *must* come down as a head as well. Clearly, the way the second coin falls is logically constrained by (i.e. connected to) how the first coin fell. These are dependent events, whatever the statistician says.'

Paradoxical in a different (and even astonishing) way is the following example from page 126 of Feller (1968). In a family with several children, define events A and B as follows. A: the family has children of both sexes, B: there is at most one girl. For a family with three children, A and B are independent, since $P(A) = 6/8$, $P(B) = 4/8$ and $P(AB) = 3/8$. But for a family with two or four children, A and B are dependent events!

By now you may be thinking that probability theory can sometimes be so counterintuitive. Indeed so – and if you browse Székely (1986), you will have further confirmation.

Questions

Question 10.1 (A)

a) We draw cards at random, with replacement, from a pack of 52 playing cards. Show that the event A: an ace is drawn, and the event B: a spade is drawn, are statistically independent events. In what way are the statements defining the events A and B logically connected?

b) What about a situation that is the converse of part (a) of this question? Can two events be statistically dependent, yet the statements that define those events be logically unconnected? If so, can you construct an illustrative example?

Question 10.2 (A)

Three coins are tossed once. What is the probability that all three coins show the same face?

If you think the answer is 0.25, how do you rebut the following argument? 'When three coins are tossed, two of them must show the same face. The third coin will show a head or a tail, in either case with probability 0.5. That means that the probability that it will show the same face as the two coins that came down alike is 0.5. So the probability that all three coins show the same face is 0.5.'

Question 10.3 (A)

a) The famous 'birthday problem' shows that if N people are gathered in the same room, the smallest value of N for which two of the people will have the same birth day and month, with specified probability, is much smaller than intuition would suggest. It is, of course, certain if there are 367 people! What is the smallest value of N for at least an even chance (probability ≥ 0.5) that two people have the same birth day and month? [Assume that births are uniformly distributed over the days of the year.]

b) Although they cannot be gathered in the same room, how many British Prime Ministers (going back from the present PM, in reverse chronological date order of first year as PM) must we consider before we find two who have the same day and month of birth?

c) Can you explain how the British Prime Ministers problem differs from the birthday problem?

Question 10.4 (B)

This question continues from the previous one.

a) Given a list of N people (whose birthdays may be considered randomly distributed through a year of 365 days), we look up their birthdays one by one in a reference book. What is the mean number of people whose birthdays have been looked up at the point where we find the first match in birthdays?

b) How do the Prime Ministers of Australia or the Presidents of the USA compare with this theoretical expectation?

Question 10.5 (B)

In 1893, a book called *Pillow Problems* was published. It contains 72 mathematical problems, of which 13 are problems in probability.

a) What was the name of the author as it appeared on the title page of the book, and what was the author's real name? In what way are the author's pen name and real name connected?

b) The Pillow Problems are stated in simple wording, but finding the correct answers is not so simple. See for yourself with this one:

A bag contains one counter, known to be either white or black. A white counter is put in, the bag shaken, and a counter drawn out, which proves to be white. What is now the chance of drawing a white counter? [Assume the bag is initially as likely to contain a white counter as a black counter.]

References

Print

Feller, W. (1968). *An Introduction to Probability Theory and Its Applications*, Volume 1, 3rd edition. Wiley.

Gigerenzer, G. (2008). *Gut Feelings: The Intelligence of the Unconscious*. Penguin.

Good, I.J. (1959). Kinds of probability. *Science* **129**, 443–447.

Székely, G. (1986). *Paradoxes in Probability Theory and Mathematical Statistics*. Reidel.

11

Some paradoxes of randomness

In CHAPTER 10, we looked at some puzzles and paradoxes of probability arising from the subtle concepts of randomness and statistical independence. We want now to uncover some paradoxes of randomness.

We saw previously that statisticians are concerned with two aspects of the concept of randomness: defining the characteristics of a random sequence of numbers (as a theoretical underpinning of the concept of a 'random variable'), and then defining operational methods for generating sequences of numbers that have these characteristics (as the basis for selecting a random sample, and for many other practical applications). The first of these aspects we can summarise by saying that a random sequence is, over a 'very long' run of numbers, notionally patternless and that, in a random sequence, each number is unpredictable from knowledge of those that came before.

We say 'notionally patternless', because there is no formal definition of a 'very long' run, and because there is no limit to the kinds of patterns we might want *not* to have in a random sequence. In practice, we have to limit ourselves to some small set of patterns to be excluded – specified so as to be appropriate to the practical context for which the random numbers are needed. We then look for a way of generating a sequence of numbers so that there is a low chance of such patterns turning up.

When it comes to unpredictability, we are similarly up against a practical constraint. In its most general conception, unpredictability requires that a specific kind of pattern be absent: there must be no exact relation connecting the nth random number to the $(n-1)^{th}$, no exact relation connecting the nth random number to any combination of the $(n-1)^{th}$ and the $(n-2)^{th}$, and so on. In practice, we usually limit ourselves to guarding against only the first of these. The discussion so far presumes that our context is one where

A Panorama of Statistics: Perspectives, Puzzles and Paradoxes in Statistics, First Edition.
Eric Sowey and Peter Petocz.
© 2017 John Wiley & Sons, Ltd. Published 2017 by John Wiley & Sons, Ltd.
Companion website: www.wiley.com/go/sowey/apanoramaofstatistics

we choose to, or are obligated to, generate (what we shall call) 'truly random' numbers. As you will see, there are statistical contexts where truly random numbers are not necessarily the statistician's first choice.

Well, then, how are truly random numbers generated and for what purposes? An everyday instance is tossing a coin to resolve an 'either-or' choice of action. Intuitively, we accept that the many small motions of the tossing hand, and chance variations in the forces applied to the coin, will ensure the unpredictability of the outcome at each toss, and an absence of systematic patterns in any long sequence of tosses. (It is assumed that there is no trickery – for example, by catching the coin and glimpsing its face before it is slapped against the wrist.) Surprisingly, however, our intuition may be wrong: coin tosses are not typically random. Such, at least, is the finding of the careful (and mathematically advanced) study of the physics of coin-tossing in Diaconis, Holmes and Montgomery (2007).

A quite different context is the public drawing of lottery prizes, where the first three prizes (say) are all substantial sums of money. Clearly, if these three prizes were won by three tickets purchased successively from the same ticket seller, questions would be asked about the integrity of the draw, *even though such an outcome is perfectly compatible with true randomness over a 'very long' run of lottery drawings.* Lottery winners must not only be chosen truly at random – they must also be *seen* to be chosen truly at random. Thus, extensive precautions are taken. For the UK National Lottery, for instance, several bins for mixing balls are available, as are several ball sets. On any particular occasion, a chance-selected bin is paired up with a chance-selected ball set. A visibly thorough mixing of the balls follows. Only then are the winning ball numbers drawn.

Despite similar precautions with all public lotteries, they are sometimes inadequate to ensure a truly random outcome. A prominent example is the 1970 US draft lottery, by which men aged between 20 and 26 were conscripted for military service in Vietnam, according to the day and month of their birth. The departures from randomness in this lottery are informatively analysed in Starr (1997), online at [11.1].

With few exceptions, it is only by 'physical' methods (i.e. methods based on a physical device, such as a lottery bin or a roulette wheel) that one can efficiently generate long sequences of the truly random numbers we have been referring to. It was, indeed, from such devices that statisticians generated random sequences in the 1930s – the early years of modern inferential statistics – for sampling and simulation studies. However, before these sequences could be (let's use the term) 'certified' as random, they needed to pass statistical tests of randomness – that is, tests suggesting the absence of several kinds of unwanted patterns. It is not surprising, then, that several tests of randomness in number sequences were devised in the same era.

Various physical methods were refined and automated in the following decades. This effort culminated in the US RAND Corporation's production, from 1949 onwards, of 'a million random digits with 100,000 normal deviates'. These million digits had been carefully tested for randomness before they were published under the above title in 1955. When this unique book of some 600 pages appeared, whimsical book reviewers enjoyed themselves ('how did they proofread it?', 'I can't recommend it: there are thousands of characters but no plot'). Today, anyone can download a copy free of charge at [11.2].

However speedily such 'certified' random sequences could be generated in that era, the process of certifying them was actually quite cumbersome, especially when extremely long sequences of random numbers were needed (e.g. for lifelike simulation of the operation of a large industrial plant). Also, all these numbers needed to be stored long-term – a challenge for the limited memory of computers of the time – because replicability of results was indispensable during the testing stage of software designed for simulation studies.

A more compact and more reliable method of generating random numbers was needed, perhaps from some kind of formula. But *what* kind of formula?

---oOo---

A remarkable formula was soon proposed – remarkable, because it is entirely deterministic! For this very reason, it is enmeshed in its own web of paradoxes. It was first convincingly demonstrated in a paper by the US mathematician D.H. Lehmer (1951). The formula Lehmer proposed is called a 'multiplicative congruential generator' (MCG). The $(n + 1)^{th}$ random integer is derived from the nth by the recurrence relation $X_{n+1} = k.X_n \pmod{M}$. Here, k and M are integer parameters: k is termed the 'multiplier' and M is termed the 'modulo'. For given values of k and X_n, X_{n+1} is the integer remainder after the integer quotient from dividing $k.X_n$ by M is evaluated. An equation involving modulo arithmetic is called a 'congruence' – hence the name of the generator.

To start the generating process off requires an integer value X_0, termed the 'seed'. Here is a simple arithmetic example: put $X_0 = 9$, $k = 11$, $M = 13$. Then $X_1 = 99 \pmod{13} = 8$, $X_2 = 88 \pmod{13} = 10$, and so on. Lehmer and dozens of subsequent writers investigated the best choices of values for k, M and X_0 to ensure that the MCG produces sequences that will pass standard tests of randomness. However, it was clear from the outset that every MCG generates sequences that are periodic – that is, after a number of steps (determinable in advance), the sequence repeats itself exactly. In the above simple example, for instance, the sequence repeats after X_{12}.

If we choose to work with an MCG, we face two paradoxes. Unpredictability, we said above, is an indispensable attribute of a random sequence. How, then, can the MCG legitimately be described as a random number generator when it is obvious that, if the formula is known, *all* the successive values it generates are entirely predictable? A sensible way to resolve this paradox is to adapt the terminology: MCG generated numbers clearly cannot be called truly random, but they can be 'pseudorandom' – that is, MCGs can (for suitable parameter values) produce sequences that will pass standard tests of randomness, even though they are not unpredictable.

Then again, how can they be essentially patternless if they are periodic? This time, we cannot escape by changing the terminology. Instead, what we must seek are parameter sets for which the MCG generates a 'certified' sequence that is long enough for our needs in a particular application, but still shorter than the period of that MCG. Studies of the period length of different MCGs are common in the scholarly literature of the past 50 years, as are similar studies for the many deterministic alternatives to the MCG formula that have been proposed as pseudorandom number generators. Today, scientists in all fields use pseudorandom numbers to solve many different kinds of statistical *and non-statistical* problems. It says a lot about the importance they attach to having suitable long-period pseudorandom number generators with excellent randomness properties that this field of research continues to be fertile, though there are already well over 1000 research papers on the theme.

Further insight on the MCG and its variants, as well as detail on some standard tests of randomness in generated sequences, can be found online at [11.3].

---oOo---

It will be clear from this discussion why winning numbers in public lotteries and in electronic gambling machines are not decided by the output of a pseudorandom number generator, but rather by truly random numbers. Truly random numbers are also preferred by many in the field of cryptography for securely encoding messages. This last avenue of application of random numbers has developed so rapidly with the growth of online commerce, mobile telephony and electronic surveillance, that there has been a renewal in recent years of research effort on ways of producing reliable truly random sequences. Old 'physical' methods (e.g. intermittent capture of mid-calculation values in computer memory; counts per unit time of atomic particles ejected from a radioactive substance) are being reappraised, and new 'physical' methods (e.g. capture of atmospheric noise; counts of the impacts of cosmic radiation) are being explored. A rare non-'physical' method has also found success, with the report that the successive digits of the decimal expansion of π form a truly random sequence – see Dodge (1966). This was no minor investigation; more than 6 billion digits of π passed multiple tests of randomness!

With so many viable sources of truly random sequences, there is now even a niche industry for the commercial supply of such sequences. A typical supplier can be found online at [11.4].

We saw above that pseudorandom number generators were devised to assure the replicability of results during the testing stage of software development, without the need to store extensive arrays of generated numbers. This was in an era when computers had sharply limited memory capacity. There is no such restriction today. Thus, it has become feasible to work extensively with generators of truly random numbers. As a philosophical bonus, it avoids the need to grapple with the paradoxical characteristics of pseudorandom number generators.

Fortunately – or unfortunately, depending on your perspective – other paradoxes of randomness remain to tantalise and disconcert us! See, for example, QUESTIONS 11.3 and 11.5, below.

Questions

Question 11.1 (A)

Seventeenth-century gamblers believed that when rolling three dice, a total of 9 and a total of 10 could each be obtained in six ways, so that rolling 9 and rolling 10 ought to have equal chances. Yet, their experience showed that 10 was more likely than 9. Who discussed and resolved this problem in an essay about the theory of dicing? And how was the problem resolved?

Question 11.2 (B)

We remarked in the Overview that multiple congruential generators (MCGs) can (for suitable parameter values) produce sequences that will pass most tests of randomness. However, there is a pattern that is unavoidable in any sequence generated by an MCG, by virtue of its recursive structure (quite apart from its periodicity). What sort of pattern is this? Investigate the pattern using the first 12 values generated by our illustrative MCG: $X_{n+1} = 11X_n$ (mod 13), with $X_0 = 9$, by plotting them, using overlapping successive pairs of these values as Cartesian coordinates.

Question 11.3 (B)

We quote from the work of a British mathematician and philosopher:

> 'We have a randomizing machine that produces a series of ones and noughts. We require for experimental purposes a random series of 16 ones and noughts. We start the machine which now gives us a series of 16 noughts. We of course reject this series as unsuitable and suspect the machine of being biased. It is returned to the makers for adjustment. When it comes back we have a very long experiment for which we require a random series of 2,000,000 ones and noughts. We leave the machine running ... but on checking through the 2,000,000 ones and noughts it produces we are surprised to find not a single run of 16 noughts. Again we suspect it of being biased and send it back. But what is its designer to say to all this? First we send it back because it produces 16 noughts in a row. Very well: he puts in a device to prevent its doing this. We then send it back because it never produces 16 noughts in a row ... It seems we are never satisfied.'

Implicit here is another paradox of the notion of randomness. What is this paradox? How do you propose that it be resolved?

Question 11.4 (B)

A traditional die has six faces, numbered 1–6, but of course it is possible to use different numberings. Here is a set of four dice which have faces numbered 1–24, with each number appearing just once:

    die A: 3, 4, 5, 20, 21, 22
    die B: 1, 2, 16, 17, 18, 19
    die C: 10, 11, 12, 13, 14, 15
    die D: 6, 7, 8, 9, 23, 24

Aside from the unusual numbering, what else is interesting about these dice, and to whom do we owe the result?

Question 11.5 (C)

Two people, A and B, play a coin-tossing game. They toss a fair coin repeatedly, with B taking over when A gets tired, and vice versa. After each toss, A scores a point if the result is a 'head', while B scores a point if the result is a 'tail'. What is the most likely value for the proportion of the time (i.e. the proportion of the total number of tosses) for which A is ahead on total points scored? If their game lasts for 1000 tosses, and A is ahead for only 50 of these, what should she conclude?

[*Note*: of course, at each toss, A is as likely to score a point as B, but something surprising happens as they keep track of their accumulating points, and thus of who is ahead. Indeed, most people would find the answers to these questions paradoxical. While the results can be stated simply, their mathematical demonstration is quite advanced!]

References

Print

Diaconis, P., Holmes, S. and Montgomery R. (2007). Dynamical bias in the coin toss. *SIAM Review* **49**, 211–235.

Dodge, Y. (1966). A natural random number generator. *International Statistical Review* **64**, 329–344.

Lehmer, D.H. (1951). Mathematical methods in large scale computing units. *Annals of the Computation Laboratory of Harvard University* **26**, 141–146.

Online

[11.1] Starr, N. (1997). Nonrandom risk: the 1970 draft lottery. *Journal of Statistics Education* **5**(2). At http://www.amstat.org/publications/jse/v5n2/datasets.starr.html

[11.2] http://www.rand.org/pubs/monograph_reports/MR1418.html

[11.3] http://www.ams.org/samplings/feature-column/fcarc-random

[11.4] http://www.random.org

12

Hidden risks for gamblers

Gambling, a passion for many people, has been around for as long as recorded history, and it shows no sign of fading away.

All gambling relates to an activity with an uncertain outcome, and with performance rules specified in advance. The activity is conventionally called a 'game', and 'gaming' is often used as a synonym (and, indeed, a euphemism) for gambling. The chief characteristic of all the games we shall look at in this chapter is that money is bet on the outcome of the game. To bet, the gambler advances a sum of money – called a 'stake' – which is the entry fee for participation in the gamble, on pre-agreed terms regarding winning and losing. After the game is played and the outcome is known, the gambler either loses the stake or regains it, together with some additional winnings. (In a few games, there is also a third possibility – see QUESTION 12.1.)

The earliest such games needed no more than the simplest equipment – bones, dice or playing cards. Making use of bleached sheep bones, the game of knucklebones (or 'jacks') was supposedly invented by a Greek soldier during the Trojan War in the 12th century BC. The oldest known dice, in use 5000 years ago, were found as part of a backgammon set at an archaeological site in Mesopotamia. Playing cards were invented in Ancient China, with early examples dating from the Tang Dynasty in the 9th century.

Today, the most commonly seen gambling devices are electronic gaming machines (called 'fruit machines' in Britain, 'slot machines' in the US and 'poker machines' in Australia). However, in our society, gambling is an inventive industry: it's hard to think of *any* activity having an uncertain outcome that does not have a gambling opportunity attached! One can, for example, bet on a horse race, a football match, share price movements, an election, or the winner of the next Oscar for best leading actor.

A Panorama of Statistics: Perspectives, Puzzles and Paradoxes in Statistics, First Edition.
Eric Sowey and Peter Petocz.
© 2017 John Wiley & Sons, Ltd. Published 2017 by John Wiley & Sons, Ltd.
Companion website: www.wiley.com/go/sowey/apanoramaofstatistics

You don't need much experience to discover that, for games devised by the commercial gambling industry, the most common result of playing the same game repeatedly is to lose all of your money. The gambler's desired outcome – to win a large sum – is far less likely than most gamblers would believe. Even when bitter experience has shown them that their desired outcome is hardly probable, compulsive gamblers persist, spurred on psychologically by irrational convictions: 'a win must come to me – because I'm special, not like others' and 'I'm owed a big win because I've had so many losses lately'. Quitting is not contemplated, because to quit would be to unequivocally accept the losses, whereas continuing leaves open the door of possibility. (See also the answer to QUESTION 3.1.)

The realities of protracted gambling have long been pointed out, sometimes quite memorably. Thus, the 16th century mathematician and inveterate gambler Girolamo Cardano wrote, 'the greatest advantage in gambling lies in not playing at all', and the first prime minister of a unified Italy, Camillo di Cavour, is reported as speaking of lotteries as 'a tax upon imbeciles'.

---oOo---

As in many other practical contexts, evaluating probabilities for gambling games can be perplexing. Even perceptive writers on this theme can be confused. Adam Smith, a pioneering Scottish moral philosopher and economist, wrote this about lotteries (in Book 1, chapter 10 of his 1776 book, *An Inquiry into the Nature and Causes of the Wealth of Nations*, online at [12.1]): 'Adventure upon all the tickets in the lottery, and you lose for certain; and the greater the number of your tickets, the nearer you approach to this certainty.' This is a puzzling assertion, explored in QUESTION 12.4.

Yet, Adam Smith correctly intuited the proposition we highlighted above: when you bet in any commercial game, you are more likely to lose than to win – and, as you continue betting, the likelihood that you will lose all your stake money approaches a certainty.

Let's investigate this proposition in detail in the context of a 'straight-up' bet at roulette – that is, a bet on a single number. A European roulette wheel is divided into 37 sectors, numbered from 0 to 36. Each sector is in the form of a shallow bin. The wheel is spun clockwise, and a small ball is tossed anticlockwise over the rotating wheel. After some bouncing around, the ball comes to rest in one of the sectors, which determines the winning number.

Suppose, at a casino, you bet $1 on the number 7. If any other number comes up, you lose your stake, and your profit is –$1; this will occur with probability 36/37 (=0.973). On the other hand, if 7 comes up, you will receive $36 and your profit is $35; this occurs with probability 1/37 (=0.027). (The fixed prize for a winning roulette bet is calculated at a rate that would be fair if there were only 36 numbers, rather than 37.) This gives a mean 'return to player' (the proportion of wagered money that is on average paid back to the player) of 0.973, or 97.3% of the stake, and a margin of 2.7% for the casino. In other words, for every $1 you bet you will, on average, get back less than $1. (By the way, European casinos are really quite modest in their margin; US casinos extract twice as much by adding another sector, labelled 00, to the roulette wheel.)

So, what happens if you play roulette repeatedly, each time making a $1 straight-up bet on the number 7? Well, most of the time, you will lose your dollar – and occasionally, roughly once every 37 spins, you will win. If you wish to work out your chances over, say, 60 such bets, you can use the binomial distribution. You are repeating your bet $n = 60$ times, the probability of success is $p = 1/37$ each time, and the results of successive spins are independent. You can calculate, for instance, the probability that you will come out ahead. For this to happen with 60 bets, you will need to win at least twice.

Formulated more generally, X, the number of wins, has a binomial (n, p) distribution. Your profit is $Y = 36X - n$, and you will be ahead if $Y > 0$, that is, if $X > n/36$. The probability is obtained from a cumulative sum of binomial probabilities. For the case of 60 bets, it is found as $1 - [(36/37)^{60} + 60(1/37)(36/37)^{59}] = 0.484$.

In this way, you can calculate your chance of coming out ahead after longer periods of betting. The third column of the table in FIGURE 12.1 shows exact probabilities for larger numbers of bets, calculated using the statistical formulae in *Excel*. We also checked them using *WolframAlpha*, online at [12.2].

Number of bets made	Minimum number of wins needed	Exact probability of being ahead	Approximate probability of being ahead
60	2	0.484	
120	4	0.408	
240	7	0.472	
480	14	0.424	
1 000	28	0.451	0.442
2 000	56	0.413	0.418
3 000	84	0.386	0.400
5 000			0.372
10 000			0.322

Figure 12.1 Exact and approximate probabilities of being ahead on roulette bets.

The probability of coming out ahead seems to be reducing as you play this game longer, though (perhaps unexpectedly) there are some increases along the way. QUESTION 12.5 explores this behaviour further.

If you were to become addicted to roulette, you might, over many years, play this game a very large number of times. In that case, to work out the chance of coming out ahead at the end, it will be arithmetically simpler to move away from the binomial distribution and invoke the Central Limit Theorem (CLT), one of the most important theorems in statistics.

---oOo---

At its most basic, the CLT says that if you draw a sample randomly from a population that is *not* normally distributed, *the sample mean will nevertheless be approximately normally distributed,* and the approximation will improve as the sample size increases.

To bring out its significance, let's amplify four aspects of this statement:

- First aspect: the CLT is about *the distribution of the mean of a random sample* from a non-normal population. What is meant by 'the distribution of the sample mean'? This is a shorthand expression for a more wordy notional concept – 'the distribution of the means of all possible samples of a fixed size, drawn from the population.' The technical term for this concept is 'the sampling distribution of the mean.'

- Second aspect: the CLT is about the distribution of the mean of a random sample from *a non-normal population.* Which non-normal population? The CLT applies to unimodal non-normal populations, whether symmetric or non-symmetric. It even applies to bimodal populations. And – using a mathematical method to match up discrete and continuous probabilities – it applies to discrete populations (such as the binomial), not just to continuous populations. Finally, it applies to empirical populations (comprising data values collected in the real world) as well as to theoretical populations (such as the binomial).

- Third aspect: the CLT says that the mean of a random sample from a non-normal population will be *approximately normally distributed* (this behaviour is sometimes called 'the CLT effect'). Does the CLT effect appear in the distribution of the sample mean from *all* populations of the kinds just listed? No – there are some theoretical populations for which the CLT effect is absent. There are examples of such populations in CHAPTER 24. A useful guide is this: if the population has a finite variance then the CLT applies. That is why the CLT certainly applies to all *empirical* populations.

(We note, in passing, that the CLT applies also to the sum of the sample values whenever it applies to the mean of the sample values. Indeed, we shall apply it below to 'total profit' from a roulette gamble. However, in all the later chapters of this book, we shall refer to the CLT only in application to the sample mean.)

- Fourth aspect: the normal approximation *will improve as the sample size increases*. How big a sample is needed for the normal approximation to look good when graphed? That depends mainly on how far from symmetry is the non-normal population from which samples are drawn. If the population is symmetric, the normal will be an excellent approximation for a sample size around 30. For a moderately skewed population, samples of size 50 will be needed. If the skewness is extreme, the CLT effect will be seen clearly only for samples of size several hundred.

The CLT effect, as the sample size increases, can be nicely illustrated graphically using one of the many CLT applets to be found on the web. A formal proof of the CLT is given in advanced textbooks.

You can see why the CLT is such a counterintuitive theorem. Who would think that the sample mean is approximately normally distributed, (almost) regardless of the form of the population being sampled? You can also now see why the CLT is such an important theorem. It *unifies* the theory of statistical inference regarding the mean. If there were no CLT, there would have to be a separate theory of inference for samples from every individual population. That would make it impossible to speak about *the* discipline of statistics.

---oOo---

Let us proceed now with calculating the probability of being ahead in roulette betting, using the CLT's normal approximation to the binomial.

As we have seen, your profit from a single straight-up bet is a binomial random variable that can take values of -1 or $+35$, with probability 36/37 and 1/37, respectively. The CLT declares that if you bet this way a very large number (N) of times, your total profit will be well approximated by a normal distribution. What are the mean and standard deviation of this normal distribution? The mean is -0.027 ($= -1(36/37) + 35(1/37)$), and it can be shown that the standard deviation is 5.838. Standard theory then tells us that the mean value of the total profit from N bets is $-0.027N$, and the standard deviation is $5.838\sqrt{N}$.

You will be ahead if your total profit is positive, so the probability of being ahead is represented by that part of the area under the normal curve that corresponds to values of profit greater than zero. For $N = 1000$ games, this gives a probability of 0.442. Results for larger numbers of games are shown in the fourth column of FIGURE 12.1. These probabilities are close to, but not the same as, the exact binomial results.

You might think that, in reality, no gambler would ever play such a large number of games of roulette. But there *are* in our society games that gamblers could well play vast numbers of times, almost without noticing. For a particularly pernicious example, consider the electronic gaming machines that we mentioned above.

In Australia, a typical poker machine might realistically have a mean return to the player of 90% (State laws require at least 85%), with a 'house' margin of 10%, a standard deviation of 17.5 (betting units) and an average electronic spin time of five seconds per play, representing 12 spins per minute. Single bets can be made in units of 1 cent or higher (frequently up to $5 and sometimes even more), and on between 1 and 50 'lines' (i.e. pre-specified winning patterns of five symbols across the screen).

Imagine a gambler who plays for a whole evening at her local club, betting 10 cents per line, 25 lines at a time, for five hours. She will play the equivalent of 90,000 games that evening at 10 cents a game, representing an outlay of $9000. Her total profit will have a mean of −9000 and a standard deviation of 5250 (in betting units of ten-cent games). In money terms, this is a mean of −$900 and a standard deviation of $525. Using the CLT approximation, her chance of being ahead at the end of the evening will be only 0.04. If she spends her evenings in this way fairly often, she will lose $900 on average each evening, and will come out ahead only on one evening out of 25.

This scenario is derived from information in government reports and from machine manufacturers; it does not represent an extreme case (see [12.3] online for one such source of information). Little wonder that social welfare groups and government agencies are worried about the effects of widespread 'problem gambling' in the community. Clubs and casinos are also worried, but seemingly more so about the possible introduction of 'gambler pre-commitment' regulations that threaten to diminish one of the richest sources of their revenue.

Unfortunately for those who gamble regularly, it seems that knowledge of probability's basic message about gambling is not widespread, nor is understanding of the power of the psychological pressures against quitting upon the gambler. If, as Cavour thought, lotteries are 'a tax upon imbeciles', then this form of high-intensity gambling is surely an even more insidious exploitation of the uninformed, the poor and the desperate.

Questions

Question 12.1 (B)

We have said that the outcome of a bet is that the gambler either loses the stake or regains it together with some additional winnings. But there is a third possibility: the gambler regains the stake, but without any additional winnings. Can you identify a casino card game where this is a possible outcome?

Question 12.2 (B)

From an ordinary deck of playing cards I take five red and five black cards, shuffle them together well, and offer you the following bet. You start with an amount of $100 (call it your 'pot'); your first bet is half of this, $50, and I match this amount. I deal out a card: if it is black, you win and take the money that we have bet ($100); if it is red, you lose and I take the money. We continue turning over the cards in the mini-deck, and each time you bet half of your current pot on the colour of the next card and I match your bet. Since there are equal numbers of black and red cards, the game will be fair. Do you agree?

Question 12.3 (A)

Which 17th century statistician and economist (predating Cavour by 200 years!) wrote in a chapter entitled 'Of Lotteries' that lotteries are an opportunity for people to tax themselves, though in the hope of gaining advantage in a particular case?

Question 12.4 (B)

Consider Adam Smith's remark: 'Adventure upon all the tickets in the lottery, and you lose for certain; and the greater the number of your tickets the nearer you approach to this certainty.' Can you suggest an interpretation of this remark that makes the statement valid or, at least, not so obviously incorrect?

Question 12.5 (C)

In the Overview, we investigated the results of repeated betting at roulette. We said that, in general, the more repeated bets you make, the lower is your probability of coming out ahead. However, FIGURE 12.1 shows that this is not uniformly true, as it indicates some increases in probability with increasing numbers of bets. This effect is, indeed, more extreme than the table implies. For a straight-up bet on a single number, most of the time simply playing one extra game will *increase* your chances of coming out ahead. Can you explain this paradox?

References

Online

[12.1] http://www.gutenberg.org/ebooks/3300

[12.2] http://www.wolframalpha.com

[12.3] Section 11.2 in chapter 11 of the *Australian Productivity Commission Report on Gambling*, Volume 1, February 2010. At http://www.pc.gov. au/inquiries/completed/gambling-2009/report/gambling-report-volume1.pdf

13

Models in statistics

'Yes, the world is a complicated place.' You have probably heard this in social conversation, and the speaker generally lets it go at that. But consider what it means for someone who is trying to understand how things actually work in this complicated world – how the brain detects patterns, how consumers respond to rises in credit card interest rates, how aeroplane wings deflect during supersonic flight, and so on. Understanding will not get very far without some initially simplified representation of whatever situation is being examined.

Such a simplified representation of reality is called a *model*. A neat definition of a model is 'a concise abstraction of reality'. It is an abstraction in the sense that it does not include every detail of reality, but only those details that are centrally relevant to the matter under investigation. It is concise in the sense that it is relatively easy to comprehend and to work with.

A simple example of a model is a street map. It shows the layout and names of streets in a certain locality and represents, by a colour coding, the relative importance of the streets as traffic arteries. It is an abstraction of reality, in that it supplies the main information that a motorist needs, but little else. For example, it is two-dimensional, and so does not show the steepness of hills, nor all the buildings that line the streets. The map is also concise in that it reduces the scale of reality to something much smaller (typically, $1\,cm = 100\,m$).

Because there are many different kinds of things in the world that we seek to understand, there are many different kinds of models. There is a basic distinction between *physical* models and *algebraic* (also called *computational*) models. A physical model is, as the name suggests, some kind of object (whether in two or in three dimensions). The circuit diagram of a

A Panorama of Statistics: Perspectives, Puzzles and Paradoxes in Statistics, First Edition.
Eric Sowey and Peter Petocz.
© 2017 John Wiley & Sons, Ltd. Published 2017 by John Wiley & Sons, Ltd.
Companion website: www.wiley.com/go/sowey/apanoramaofstatistics

digital radio receiver is evidently a physical model. So is an architect's three-dimensional representation of a house as it will look when built, and so also is a child's toy helicopter.

An algebraic model, on the other hand, uses equations to describe the main features of interest, as well as their interrelations, in a real-world situation. If these equations describe relations that are certain, or relations where chance influences are ignored, the model is called a *deterministic* model or (alternatively) a *mathematical* model. Newton's three 'laws' of motion and Einstein's famous equation $E = mc^2$ are examples of mathematical models.

If, however, the equations explicitly include the influence of chance, then the model is called a *stochastic* model or (alternatively) a *statistical* model. Thus, a statistical model, by definition, always includes a purely random (i.e. *completely unpredictable*) component, to represent the influence of chance upon the quantity being modelled.

A statistical model may additionally include a deterministic component, comprising one or more distinct systematic (i.e. *either known or predictable*) influences upon the quantity being modelled. We shall look at such statistical models later in this Overview.

We find it convenient here to refer to a statistical model which has no deterministic component as a *probability model*. Though introductory textbooks of statistics may not highlight the fact, all the standard probability distributions (binomial, Poisson, normal, etc.) are, indeed, probability models.

---oOo---

To see the binomial distribution as a probability model, let's take a real-world situation. A big bin of apricots is delivered and someone pulls out four apricots for us to eat. What is the probability (we may be interested to know) of getting three ripe apricots and one unripe apricot in a selection of four apricots? It is difficult to answer this question in the real world, because there are so many things about this situation that we don't know. For example, we don't know (a) whether the apricots were selected deliberately or at random, (b) how, exactly, to tell a ripe apricot from an unripe one, and (c) how the ripe and unripe apricots are distributed through the bin.

We can simplify the problem if we make some assumptions. From the textbook, we learn that (i) if we select the apricots *at random*; and (ii) if we define a selected apricot as either ripe or unripe (that is, *there are only two possible outcomes*); and (iii) if the chance of selecting a ripe apricot from the bin *is always the same*, each time we select an apricot; and (iv) if the occasions on which we select an apricot are unconnected with (that is, *independent* of) each other, then there is a concise formula – the binomial probability density function – for evaluating the probability of getting three ripe apricots and one unripe apricot in a selection of four apricots from the bin.

These four assumptions collectively imply a quite marked abstraction from the reality of the situation. Let's see how. Firstly, there could be a third possible outcome: a selected apricot might look ripe, but be unripe inside. Secondly, once most of the apricots in the bin have been drawn out, it will not be very realistic to continue to assume that that the chance of selecting a ripe apricot is constant at each draw. After all, every time an unripe apricot is selected from the bin, that increases the chance that a ripe apricot will be selected from the remaining apricots at the next draw. Finally, selections – even if they are seemingly random – will not always be independent in practice. If an apricot is drawn for eating, but turns out to be unripe, another apricot will be immediately selected. That subsequent draw will occur precisely because the previously drawn apricot was unripe. That is *not* independence of draws!

Before analytical use is made of any statistical model – and that, of course, includes a probability model – it is clearly important to *validate* the model. This involves checking that the model captures all the characteristics of the real world that are essential for the purpose at hand. Checking the 'fit' of the model to the real world is a two-stage process: ensuring that the model is *practically suitable*; and then testing that the model's predictions of real-world data are a *statistically close* match to the actual real-world data. Good applied statistical work requires careful attention to both stages.

Returning to our apricots example, if the four assumptions set out above are judged sufficiently realistic, then the binomial probability density function (for short, say simply 'the binomial distribution') will be a practically suitable model. On the other hand, if we judge it essential to take into account the possibility of three outcomes (apricot ripe, apricot unripe, or apricot looks ripe but is unripe), then a practically suitable model will be the *trinomial distribution*. Alternatively, if two possible outcomes are sufficiently realistic, but we are not comfortable with assuming that the probability of drawing a ripe apricot remains the same each time we select an apricot, then a practically suitable model will be the *hypergeometric distribution*. You may not be familiar with these standard probability distributions – the trinomial and the hypergeometric – but that need not be an obstacle to following this discussion.

The point is that when some standard probability model is not practically suitable, there is often another one already in the statistician's tool kit that fits the bill more appropriately. However, if there is not, then it is usually possible to develop an improved model from scratch – see Gelman and Nolan (2002), pages 142–145, for an instructive example in the context of putting in golf.

Suppose we decide to adopt the binomial distribution as a practically suitable model for finding the probability of x ripe apricots in a random draw of four apricots, where x runs from 0 to 4. Next, we have to test whether the model's predictions of real-world data are a *statistically close* fit to those real-world data. To generate some real-world observations, we perform, say, 120 random draws of 4 apricots. Suppose we find:

No. of ripe apricots per draw	4	3	2	1	0
Observed no. of draws	10	30	42	26	12

The binomial distribution model defines the probability of x ripe apricots in a random draw of four apricots by the formula $^4C_x\, p^x\, (1-p)^{4-x}$. To proceed, we need to estimate the parameter p, the probability of selecting a ripe apricot at any one draw. We note the property of the binomial distribution model that the mean number of ripe apricots in selections of four apricots is $4p$. Next, we calculate the weighted mean number of ripe apricots per draw from the above real-world data. This is 2.0. Equating the model's and the real-world data's mean values, $4p^* = 2.0$, we find an estimate of p to be $p^* = 0.5$. We can now calculate model-predicted probabilities, and frequencies over 120 draws:

No. of ripe apricots per draw	4	3	2	1	0
Predicted probability	0.0625	0.25	0.375	0.25	0.0625
Predicted no. of draws	7.5	30	45	30	7.5

The statistical fit of the binomial distribution model can then be tested by the chi-squared goodness-of-fit test. The test statistic is $K = \Sigma[(O_i - P_i)^2 / P_i]$, where O is the observed and P the predicted number of draws, and $i = 1, 2, \ldots$ 5. Small values of K (implying O and P values close together) signal a good fit of the data to the model. The critical value of the chi-squared statistic, with (here) 3 degrees of freedom and a 5% level of significance is 7.81. This means that if $K > 7.81$, there is a statistically significant difference between the model's predictions and the real-world observations – that is, the test suggests the data are a poor fit to the model. In other words, the model is a poor one.

However, if $K \leq 7.81$, the model's predictions do not contradict the data, so we may act as if the model is a good one. For these data, $K = 4.27$, so we conclude (with a 5% risk of a type I error) that probabilities in the real world are, statistically, a close fit to the model. We may, thus, use the binomial distribution model with some confidence in further analyses that may be of interest in connection with this crop of apricots – for example, to infer from a sample whether another bin of these apricots contains, on delivery, an unacceptable number of unripe ones.

This has been an example of validation of a univariate probability model – a univariate statistical model with no deterministic component.

---oOo---

Statistics is also concerned with bivariate models (e.g. the simple regression model) and higher-order (multivariate) models. Let's look now at a bivariate relation, and contrast the way it is represented by a mathematical modeller and by a statistical modeller. An object, initially at rest, is struck by a momentary force. The object moves (in a straight line) a distance d (metres) in elapsed time t (seconds). The relation is that between d and t.

How does the mathematical modeller proceed to model this bivariate relation?

A plot of d against t, using n pairs of experimental (t, d) data, suggests that d *approximates* some curvilinear function of t. Guided by a theory in physics (i.e. one of Newton's laws of the motion of a body moving from rest in a straight line), the modeller chooses the quadratic function $d = bt^2$, where b is some constant, as a practically suitable deterministic model. It should be noted that a practically suitable model generally has a firm foundation in accepted theory.

A value of b is found by using some criterion of 'best-fit' of the data to the model, say, the criterion of least squares. Thus, minimising the value of $\sum_1^n (d_i - bt_i^2)^2$, the best-fit value of b is calculated. Suppose it is 3.72. The modeller then draws the graph of $d = 3.72\,t^2$. If this graph looks (to the eye) pretty close to the plot of the experimental data (a rather informal criterion!), then this deterministic model is considered validated.

Notice that, although the plot shows that the experimental data do not lie *exactly* along the quadratic curve $d = 3.72\,t^2$, the mathematical modeller does not model the discrepancies. The entire focus is on the deterministic function $d = bt^2$.

What about the statistical modeller?

The statistical modeller begins in the same way as the mathematical modeller, by plotting the n pairs of experimental (t, d) data and (from a theory in physics) settling on the practically suitable deterministic function of the form $d = bt^2$. At this point, the procedures diverge.

The statistical modeller sees that the experimental data do not lie *exactly* along a quadratic curve, and asks what could be the reason for the discrepancies. An immediate thought is that not all the influences that govern d have been taken into account. For instance, distance travelled will also depend on the nature of the surface over which the object is travelling – in other words, on the amount of friction. It will also depend on the mass and shape of the object. Perhaps there is also wind assistance to the object's travel. Further reflection might add to this list of influences – all of them ignored in Newton's statement of his laws of motion, but all relevant to making the model more realistic.

The statistical modeller could attempt to measure individually each of these influences that disturb (or displace) the experimental values from the deterministic function $d = bt^2$, but this would certainly complicate the analysis. At the cost of some realism, it is open to the modeller to lump all the additional influences together and to refer to their net value, for a given value of t, as *the* disturbance corresponding to that value of t.

Denote the disturbance by the symbol e. Since each value of e is, in principle, the net value of a large number of independent influences – some slowing the object down and some speeding it up – e behaves like a random variable. The statistical model is then: $d = bt^2 + e$

Notice that this model has a deterministic component, involving the systematic variable t, and a random component e.

In this example, it is realistic that the two components are added together, but in other contexts it may be more appropriate to combine them in another way – say, by multiplying them together. To make this statistical model operational, it is necessary to choose a probability model for e. Here, it is also realistic to use the normal distribution to model the values of e.

The mathematical modeller writes $d = bt^2$, suppressing the random disturbance. What does the statistical modeller gain over the mathematical modeller by explicitly modelling the disturbance? It is the possibility of *formally evaluating* the statistical model as an accurate representation of reality. There is no such possibility for the mathematical model.

The following steps of reasoning make clear what we mean by formal evaluation in this context.

Firstly, the statistician identifies the concept of a sample estimate of a population parameter, and highlights its important role in the modelling of observed data.

When the mathematical modeller finds the value 3.72 by least squares, based on the single sample of (t_i, d_i) pairs, $i = 1, 2, 3 \ldots n$, this value is treated as a fixed *constant*. And there the formal part of the modelling procedure concludes.

However, when the statistical modeller finds the value 3.72 by least squares, based on the single sample of (t_i, d_i) pairs, $i = 1, 2, 3 \ldots n$, it is with the insight that the least-squares estimator (the formula that generates a numerical estimate for any set of sample data) could produce values somewhat different from 3.72 when applied to other samples of n pairs of (t_i, d_i) data.

Let's use $\hat{b}$ to represent the least-squares estimator (formula). It follows that $\hat{b}$ is a *variable*. Each value of $\hat{b}$ is a sample estimate of the unknown population parameter, b.

Secondly, the statistician recognises that $\hat{b}$ is a *random* variable (with a specifiable form of probability distribution). This follows because $\hat{b}$ can (easily) be shown to be a function of the random variable, e. The statistical distribution of the estimator $\hat{b}$ is a transformation of the probability distribution used to model e.

Given the statistical distribution of $\hat{b}$, a test of statistical significance can be derived for formally evaluating the accuracy of the statistical model as a representation of reality.

To summarise, whether a statistical model is composed of a deterministic component and a random component, or of a random component alone, the procedure for validating the model is the same: first ensure that the model is practically suitable, then test *formally* that its predictions are statistically close to what is observed in the real world.

The British statistician George Box went straight to the heart of the matter when he wrote 'all models are wrong but some are useful' (see Box, 1979). With a valid and, therefore, useful statistical model, it may then become a little easier to understand how it works – this complicated place, the world.

Questions

Question 13.1 (A)

When a fair coin is tossed, we are accustomed to writing P(Head) = 0.5 and P(Tail) = 0.5. In what other ways could the coin toss turn out? What probability, then, is being assigned to those other outcomes? What conclusion do you draw from this exploration?

Question 13.2 (A)

Under what circumstances will the normal probability distribution be a practically suitable statistical model for the heights of adult humans?

Question 13.3 (B)

Which probability model was originally developed from a consideration of the deliberations of juries in the early 19th century, and found to be of immense value in the mid-20th century, in relation to the German V2 rockets bombarding London towards the end of World War II?

Question 13.4 (B)

a) The exponential distribution is widely used as a probability model in the analysis of queuing problems. What particular property does it have that makes it useful in such contexts?
b) A tourist information office is staffed by two officers, and the time they spend serving any customer is exponentially distributed, with the same mean service time for each. When C walks into the office, she finds A and B each being served by one of the officers. As soon as an officer is free, she will be served. What is the probability that C is the last of the three customers to leave the office?

Question 13.5 (B)

When arrival probabilities in queuing problems are modelled by the Poisson distribution, the probability of two arrivals in two hours is not equal to twice the probability of one arrival in one hour; in fact, it is *considerably less* than twice the probability of one arrival in one hour. Can you give a straightforward and intuitively satisfying reason why this is so?

References

Print

Box, G.E. (1979). Robustness in the strategy of scientific model building. In: Launer, R.L. and Wilkinson, G.N. (eds). *Robustness in Statistics*, pp 201–236. Academic Press.
Gelman, A. and Nolan, D. (2002). *Teaching Statistics: A Bag of Tricks*. Oxford University Press.

14

The normal distribution: history, computation and curiosities

When we began our statistical studies at university, last century, we bought, as instructed, a booklet of 'Statistical Tables'. This booklet contained all the standard tables needed for a conventional undergraduate degree programme in statistics, including the 'percentage points' of a variety of standard distributions, critical values for various statistical tests, and a large array of random numbers for sampling studies. We were soon made aware of one particular table, titled 'Areas under the standard normal curve', and were left in no doubt that we would be referring to it frequently. We used this booklet of tables in classwork throughout our studies, and had clean copies of it issued to us at every statistics examination. From our vantage point today, that booklet of Statistical Tables has become a rather quaint historical artefact from the mid-20th century, at a time when calculators were the size of typewriters (both of these, too, being artefacts of that era).

That vital standard normal area table was to be found also in the Appendix of every statistics textbook on the market at that time. That is still the case today. It suggests that this printed tabulation from the past is still being consulted by students and, perhaps, also by professional statisticians. Need this still be so?

---oOo---

Before considering this question, let's look at the history of the normal distribution and of the construction of this enduring table.

The normal distribution of values of the variable x has probability density function (pdf):

$$f(x \mid \mu,\sigma) = \frac{1}{\sigma\sqrt{2\pi}} \exp\left(-\frac{1}{2}\left(\frac{x-\mu}{\sigma}\right)^2\right)$$

A Panorama of Statistics: Perspectives, Puzzles and Paradoxes in Statistics, First Edition.
Eric Sowey and Peter Petocz.
© 2017 John Wiley & Sons, Ltd. Published 2017 by John Wiley & Sons, Ltd.
Companion website: www.wiley.com/go/sowey/apanoramaofstatistics

with parameters μ (the population mean) and σ (the population standard deviation). It is the most commonly met probability distribution in theoretical statistics, because it appears in such a variety of important contexts. If we follow the historical evolution of this distribution through the 18th and 19th centuries, we shall discover these contexts in a (somewhat simplified) sequence.

The normal distribution (though not yet with that name) first appeared in 1733, in the work of the English mathematician Abraham de Moivre. De Moivre discovered that the limiting form of the (discrete) binomial distribution, when the number of trials of the binomial experiment becomes infinitely large, is the pdf that we today call the (continuous) normal distribution. Stigler (1986), pages 70–85, conveys, in some analytical detail, the satisfaction de Moivre gained from his hard-won discovery.

In a book on planetary motion, published in 1809, the German mathematician Carl Friedrich Gauss presented his pioneering work on the method of least squares estimation. It was in this context that he proposed the normal distribution as a good theoretical model for the probability distribution of real-world random errors of measurement. Now, in a context different from de Moivre's, Gauss rediscovered the pdf of the normal distribution by asking the question (translated into modern terminology): for what symmetric continuous probability distribution is the mean of a random sample the maximum likelihood estimator of the population mean? How he derived the pdf algebraically is sketched in Stigler (1986), pages 139–143.

It's worth mentioning here that Gauss's choice of a model for random errors of measurement was not the only candidate historically considered for that purpose. Gauss's contemporary, the French mathematician Pierre Simon Laplace, had independently been looking for a suitable model since about 1773. However, he tackled the problem in a way that was the converse of Gauss's. Gauss simply sought the symmetric function whose mean was 'best' estimated by the sample mean. Happily, he hit upon a criterion of a 'best' estimator that produced a model with other far-reaching virtues as well.

Laplace's approach was to search ingeniously among mathematical functions having the desired graphical profile – unimodal, symmetric and with rapidly declining tails – and to worry afterwards about how the parameters would be estimated. During the following dozen years, he came up with several promising models, but they all ultimately proved mathematically intractable when it came to parameter estimation. Some of the candidate functions that Laplace investigated are on view in Stigler (1986), chapter 3.

Though Laplace was familiar with the pdf of the normal distribution from de Moivre's work, he somehow never thought of considering it as a model for random measurement errors and, after 1785, he turned his attention to other topics.

Then, in early 1810 – shortly after Gauss's rediscovery of the normal distribution in 1809 – Laplace found the normal distribution turning up in his own work. This was, again, in a different context – his early proof of what we know today (in a more general form) as the Central Limit Theorem (CLT). You will find a statement of the CLT in the Overview of CHAPTER 12. To recapitulate: under very general conditions, the mean of a sample from a non-normal distribution is approximately normally distributed if the sample size is large.

These several important achievements prompted many 19th century mathematicians to call the normal the Gauss-Laplace distribution. As time passed, however, that name gave way to simply the Gaussian distribution.

The name 'normal' for this distribution first appeared when Francis Galton pioneered it to a wide public in his 1889 book *Natural Inheritance* (online at [14.1]) – giving chapter 5 the title 'Normal Variability', while still occasionally using an earlier name, 'law of frequency of error'. He and many of his scientific contemporaries were excited to confirm that biologically determined real-world variables, such as species-specific size and weight, are often approximately normally distributed. Thus, yet a fourth context for the significance of this distribution was identified. This last context is, logically, quite closely related to the normal as a model for random errors of measurement, as Gauss had earlier proposed.

Galton was so moved by his and his predecessors' discoveries that he imbued the normal distribution with an almost mystical character (see QUESTION 14.3). He writes (chapter 5, page 66): 'I know of scarcely anything so apt to impress the imagination as the wonderful form of cosmic order expressed by the "Law of Frequency of Error". The law would have been personified by the Greeks and deified, if they had known of it. It reigns with serenity and in complete self-effacement amidst the wildest confusion. The huger the mob, and the greater the apparent anarchy, the more perfect is its sway.' (See QUESTION 22.1(b) for a 20th century tribute in only slightly less lyrical terms.)

From Galton's time onwards, English-speaking statisticians began to use the term 'normal distribution' routinely. Continental statisticians, on the other hand, continued for many years to refer to it as the 'Gaussian distribution'.

For broader historical detail on this major strand in the history of statistics, we recommend the introductory chapter of Patel and Read (1996).

---oOo---

What about the calculation of normal probabilities? As for any continuous probability distribution, (standard) normal probabilities are represented by areas under the (standard) normal curve. The equation of the standard normal curve is $f(x) = (1 / \sqrt{(2\pi)})\exp(-\frac{1}{2}x^2)$. There is a practical obstacle to evaluating areas under this curve by straightforward application of integral calculus. This is because (as has long been known) there is no closed-form expression for the indefinite integral of the right hand side function in the above equation – that is, no solution in terms only of constants; variables raised to real powers; trigonometric, exponential or logarithmic functions; and the four basic operators $+ - \times \div$.

Happily, there are several alternative algorithms for numerically approximating the definite integral between any two given values of x, and thus finding the corresponding normal area to any desired degree of accuracy. In earlier times, such a calculation would be slowly and laboriously done by hand, with much checking of numerical accuracy along the way. Nowadays, numerical approximation software does the job swiftly and effortlessly.

Fortunately, the standard normal area table suffices for evaluating areas under any normal curve, since all normal curves have the same shape, relative to their location and spread. A random variable that has a normal distribution with a general mean μ and general standard deviation σ can be standardised (i.e. transformed into a standard normal random variable, having mean 0 and standard deviation 1) by subtracting μ and dividing by σ.

Normal area tables were first calculated in the late 18th century, and for a variety of purposes. In 1770–71 in Basel, Daniel Bernoulli compiled a table of areas under the function $y = \exp(-x^2/100)$, essentially a multiple of a normal distribution function, for approximating binomial probabilities. In 1799 in Strasbourg, Chrétien Kramp prepared a similar table to aid astronomical calculations of refraction. An overview of normal area tables published between 1786 and 1942, with insights on the different algorithms used, is given in David (2005). It is interesting to note that the normal area tables produced in 1903 by the Australian-English statistician William Sheppard (1863–1936) have remained unsurpassed in scope and accuracy, with later reproductions of these tables differing mainly in their layout.

---oOo---

Do students of statistics still need booklets of statistical tables?

There is no doubt that they were indispensable until the microcomputer (also called a personal computer or desktop computer) became ubiquitous about 35 years ago. Thereafter, they became merely convenient, but even that is hardly the case today. We can see why by tracing the evolutionary thread of computing devices over the past century. (We are omitting mention of mainframe and minicomputers, because these were not generally available to undergraduate statistics students.)

During this period, the time interval between successive significant innovations in computing technology has been shortening, the computing power of the devices has been growing, and their physical scale has been shrinking – from the 'transportable', to the 'portable', to the 'mobile'. For the practising statistician, complex computation without a mechanical calculator was unthinkable in the period 1900–1940. Electrically driven calculators gradually took over in the years 1940–1970. These were, in turn, superseded by hand-held electronic calculators (first solely hard-wired, then later programmable) over the period 1970–1985. These devices, with further refinements such as graphics capability, then co-existed with the evolution of computationally much more powerful, but bulkier, microcomputers (from around 1970) and laptop computers (from around 1990). Smaller-sized netbooks emerged in 2007, and tablet computers in 2010. Today, mobile 'smart' phones and (even smaller) 'wearable' devices represent the latest reductions in scale.

From this overview, we see that it was only with the arrival of programmable calculators, around 1980, that devices powerful enough for automated statistical computation first became cheap enough for students to afford one for their own personal use.

Today, web-enabled and app-equipped tablets and phones have comprehensively displaced programmable calculators for the standard repertoire of statistical functions and analyses. With steady growth in the range of statistical apps being made available, and seemingly endless expansion of statistical resources (including applets) on the web, these highly mobile personal devices can be very efficient tools for routine statistical computing, including finding 'areas under the normal curve'.

Moreover, applets and apps introduce two improvements over printed standard normal area tables. The first is direct computation of any area under any normal distribution, using the option of specifying the mean and standard deviation of the required normal distribution. The second is graphical representation of the area calculated – that is, the probability. The first offers only convenient flexibility, but the second is an important aid for students learning about the normal distribution and finding normal probabilities for the first time. As teachers, we know that this step in a first course in statistics is often a real stumbling block; visual representation in this, and many other contexts, can be a significant key to learning.

Here are two examples of these types of software. David Lane's online textbook of statistics, HyperStat [14.2] includes an applet for finding the normal area corresponding to the value of a standard normal variable (or vice versa) [14.3]. A similar function is available in the modestly-priced StatsMate app (see [14.4]), developed by Nics Theerakarn for a tablet computer or mobile phone. Both sources also provide routines for a wide array of other statistical calculations, as their websites reveal.

We recognise that the limited availability and high expense of modern mobile technology in some countries may preclude access to the convenience this technology offers. For those to whom it is available and affordable, however, the era of statistical tables is, surely, past.

Questions

Question 14.1 (A)

Attributes of the normal distribution, some of which may surprise you.

a) As everyone knows, the standard normal distribution has points of inflection at $z = -1$ and $+1$. But where do the tangents at these points cut the z-axis?
b) On the standard normal distribution, what is interesting about the z-value, $z = 0.35958$?
c) Suppose we wish to draw a standard normal distribution accurately to scale on paper so that the curve will be 1 mm above the horizontal axis at $z = 6$. How large a piece of paper will be required? [*Hint*: how high will the curve be above the horizontal axis at the mode?]

Question 14.2 (B)

William Sheppard's (1903) article, 'New tables of the probability integral', gives cumulative areas under the standard normal curve to 7 decimal places for values of z from 0.00 to 4.50 and to 10 decimal places for values of z from 4.50 to 6.00. Why did he carry out his calculations to so many decimal places? Can you suggest any situation where knowledge of these values to such accuracy would be useful?

Question 14.3 (B)

In the Overview, we quoted Galton's lyrical description of the normal distribution as that 'wonderful form of cosmic order'. Galton's jubilation came from observing two phenomena. Firstly, that the distribution of *measured values* of variables which could be interpreted, in some way, as random errors (i.e. deviations from some biological or technical standard, or 'norm'), seems to be (approximately) normal. And secondly, that the distribution of *sample means* drawn from non-normal distributions becomes, as the sample size increases ('the huger the mob'), more and more like the normal ('the more perfect its sway') – which is what the Central Limit Theorem (CLT) declares.

However, there are exceptions – measured variables with distributions that do not look normal, and sampling distributions of sample means that do not conform to the CLT. Give examples of these two kinds of exceptions. Can they be brought within the 'cosmic order'? Do you think Galton was wildly overstating his case?

Question 14.4 (B)

Some non-normal data have a distribution that looks more like a normal after a logarithmic transformation is applied. Other non-normal data look more like a normal after a reciprocal transformation is applied. What data characteristic(s), in each case, suggest that the mentioned transformation will be effective?

Question 14.5 (B)

Sketch on the same set of axes the frequency curve of a standard normal distribution and of a chi-squared distribution with one degree of freedom. Do these two curves intersect? (*Note*: the chi-squared distribution with one degree of freedom is the distribution of the square of a single standard normally-distributed variable.)

References

Print

David, H.A. (2005). Tables related to the normal distribution: a short history. *The American Statistician* **59**, 309–311.

Patel, J.K. and Read, C.B. (1996). *Handbook of the Normal Distribution*, 2nd edition. CRC Press.

Sheppard, W. (1903). New tables of the probability integral. *Biometrika* **2**, 174–190.

Stigler, S.M. (1986). *The History of Statistics: The Measurement of Uncertainty Before 1900*. Harvard University Press.

Online

[14.1] Galton, F. (1889). *Natural Inheritance*. Macmillan. At http://galton.org, click on Collected Works, then on Books.

[14.2] http://davidmlane.com/hyperstat/index.html

[14.3] http://davidmlane.com/normal.html

[14.4] http://www.statsmate.com

Part IV

Statistical inference

15

The pillars of applied statistics I – estimation

There are two pillars of statistical theory, upon which all applied work in statistical inference rests. In this chapter we shall focus on estimation while, in the next chapter, we shall look at hypothesis testing. Among the most famous of past statisticians, Ronald Fisher, Jerzy Neyman and Egon Pearson (whose names appear in FIGURE 22.2) laid the foundations of modern methods of statistical inference in the 1920s and 1930s. They polished procedures for estimation proposed by earlier thinkers, and invented terminology and methods of their own. This was an era of fast-moving developments in statistical theory.

For relevant historical background, a valuable resource is Jeff Miller's website at [15.1], titled *Earliest Known Uses of Some of the Words of Mathematics*. The entry for 'Estimation' informs us that the terms 'estimation' and 'estimate', together with three criteria for defining a good estimator – 'consistency', 'efficiency' and 'sufficiency' – were first used by Fisher (1922), online at [15.2]. Fisher defined the field in a way that sounds quite familiar to us today: 'Problems of estimation are those in which it is required to estimate the value of one or more of the population parameters from a random sample of the population.' In the same article, he presented 'maximum likelihood' as a method of (point) estimation with some very desirable statistical properties.

Neyman, who subsequently pioneered the technique of interval estimation, referred to it as 'estimation by interval', and used the term 'estimation by unique estimate' for what we now call point estimation. It was Pearson who introduced the modern expression 'interval estimation'.

These, and many earlier, pioneers of (so-called) Classical estimation devised parameter estimators that use sample data alone. In addition to

A Panorama of Statistics: Perspectives, Puzzles and Paradoxes in Statistics, First Edition.
Eric Sowey and Peter Petocz.
© 2017 John Wiley & Sons, Ltd. Published 2017 by John Wiley & Sons, Ltd.
Companion website: www.wiley.com/go/sowey/apanoramaofstatistics

maximum likelihood, Classical techniques include method of moments, least squares, and minimum-variance unbiased estimation.

The pioneers' successors pursued more complex challenges.

A salient example is devising estimators that are more efficient than Classical estimators, because they synthesise the *objective* (i.e. factual) information in sample data, with additional *subjective* information (e.g. fuzzy knowledge about parameter magnitudes) that may be available from other sources. For example, if we want to estimate the mean of a Poisson model of the distribution of children per family in Australia, we can expect to do so more efficiently by amending a Classical estimation formula to incorporate the 'common knowledge' that the mean number of children per family in Australia is somewhere between 1 and 3.

The most extensive array of techniques for pooling objective and subjective information to enhance the quality of estimation goes by the collective name *Bayesian estimation*. Methods of Bayesian estimation were first proposed in the 1950s, by the US statistician Leonard Savage. See CHAPTER 20 for an overview of the Bayesian approach.

Here is a second example of progress in estimation.

The Classical theory of interval estimation of any parameter – say, a population mean – *necessarily assumes a specific model* (e.g. normal, exponential, Poisson) for the distribution of the population data. As you may know, constructing a confidence interval for a population mean requires a value both for the sample mean (or whichever other point estimator of the population mean is to be used) and for the standard error of the sample mean (or other estimator). Often, the theoretical standard error involves unknown parameters, so it is necessary, in practice, to work with an estimated standard error. The Classical approach uses the properties of the specific model chosen for the data to derive an estimator of the standard error.

But how is such a model fixed upon in the first place? It may be suggested by theoretical principles of the field of knowledge within which the data are being analysed, or by the general appearance of the summarised sample data (e.g. a histogram or a scatter diagram).

Yet, what if the field of knowledge has nothing to say on the choice of a model for the data – and what if, moreover, the summarised sample data look quite unlike any of the well-established statistical models? We would then need to find some way of estimating the standard error of the sample mean *without having any explicit model* as a basis. In 1979, the US statistician Bradley Efron invented a very effective way of doing just that. His estimation procedure is known as bootstrapping. In the simplest version of this procedure, the sampling distribution of the sample mean is approximated by repeated sampling (termed 'resampling') from the original sample.

A bootstrapped standard error of the sample mean can be constructed from this distribution and, thus, a bootstrapped confidence interval for the population mean can be obtained. For a fuller, non-technical explanation (including an illustration of resampling), see Wood (2004) or Diaconis and Efron (1983).

Given the complexity of the real world, and the ever-increasing number of fields in which statistical methods are being applied, it is hardly surprising that countless situations have turned up where no well-defined model for the data is evident, or where statisticians are unwilling to assume one. This explains the enormous growth in popularity of estimation by bootstrapping over the past twenty years.

Lastly, a third direction in which estimation has moved in the post-Classical period: statisticians' willingness to use biased estimators.

In Classical estimation, whenever it came to a conflict between the criteria of unbiasedness and efficiency in the choice of a 'best' estimator, the unbiased estimator was inflexibly preferred over the more efficient, but biased, one. QUESTION 6.5 illustrates such a conflict.

A more flexible way of resolving this kind of conflict is to see if there is an estimator that compromises between the conflicting criteria – that is, an estimator which is a little biased, yet rather more efficient than the corresponding unbiased estimator. There are several paths to finding such useful biased estimators. One is the method of minimum mean square error (MMSE) estimation. A context in which this is effective is seen in QUESTION 15.4.

Unfortunately, the method of MMSE estimation is not immune to breakdown, even in some quite simple contexts. See QUESTION 15.5(a) for an example.

Classical estimation methods break down, too, though more rarely. Maximum likelihood estimation, for instance, fails in any context where the likelihood function increases without limit, and thus has no maximum. A (quite technical) example is given in Konijn (1963).

You have read of paradoxes in earlier chapters. So it should come as no surprise that there are paradoxes to be found (and resolved!) in the theory of estimation as well. There are two in the answer to QUESTION 15.5 (b).

Questions

Question 15.1 (B)

Suppose you are asked this question: 'I've noticed that the best estimator of the population mean is the sample mean; the best estimator of the population median is the sample median; and the best estimator of the population variance is the sample variance. Is that a pattern I can rely on for finding best estimators?' How would you answer?

Question 15.2 (B)

Given a sample mean and an initial numerical 95% confidence interval for the unknown population mean of a normal distribution, $N(\mu, \sigma^2)$, based on that sample mean, what is the probability that a replication (i.e. an independent repetition of the same sampling process) gives a sample mean that falls within the initial confidence interval? [For simplicity, assume that the value of σ^2 is known.]

Question 15.3 (B)

We wish to estimate the mean μ of a normal distribution $N(\mu, \sigma^2)$. Suppose we have two independent random samples from this distribution: one sample has size n_1 and sample mean $\bar{X}_1$, and the other has size n_2 and sample mean $\bar{X}_2$. As an estimator of μ, is it better to use the average of the sample means $\frac{1}{2}(\bar{X}_1 + \bar{X}_2)$ or, alternatively, the mean of the pooled data $[1 / (n_1 + n_2)] \left[\sum_{i=1}^{n_1 + n_2} X_i \right]$?

Question 15.4 (C)

When estimating the variance σ^2 of a normal population with unknown mean from a sample of size n with mean $\bar{X}$, we know (see, for example, QUESTION 6.5) that $\Sigma(X - \bar{X})^2 / (n-1)$ is an unbiased estimator. But what sort of *biased* estimator is $\Sigma(X - \bar{X})^2 / (n+1)$, and why might we prefer to use it?

Question 15.5 (C)

a) In the case of the normal distribution $N(\mu, \sigma^2)$, with σ^2 assumed known, consider estimators of μ of the form $c\,\bar{X}$ (c a constant), where $\bar{X}$ is the sample mean. Find the value of c that will make $c\,\bar{X}$ the MMSE estimator of μ, and show that this value of c means that here the method of MMSE estimation has failed.

b) In 1961, in the *Proceedings of the 4th Berkeley Symposium on Mathematical Statistics and Probability*, two US statisticians, Willard James and Charles Stein, published a very counterintuitive – and, indeed, paradoxical – theoretical result loosely to do with MMSE estimation. What is this result? And why is it paradoxical?

References

Print

Diaconis, P. and Efron, B. (1983). Computer-intensive methods in statistics. *Scientific American* **248**(5), 116–130.

Konijn, H.S. (1963). Note on the non-existence of a maximum likelihood estimate. *Australian Journal of Statistics* **5**, 143–146.

Wood, M. (2004). Statistical inference using bootstrap confidence intervals. *Significance* **1**, 180–182.

Online

[15.1] http://jeff560.tripod.com/mathword.html

[15.2] Fisher, R.A. (1922). On the mathematical foundations of theoretical statistics. *Philosophical Transactions of the Royal Society, Series A* **222**, 309–368. At http://digital.library.adelaide.edu.au/dspace/handle/2440/15172

16

The pillars of applied statistics II – hypothesis testing

In this chapter, we are looking at hypothesis testing – that peculiarly statistical way of deciding things. Our focus is on some philosophical foundations of hypothesis testing principles in the frequentist, rather than the Bayesian, framework. For more on the Bayesian framework, see CHAPTER 20.

The issues discussed here all relate to a single test. In the next chapter, we investigate some matters that may complicate the interpretation of test results when multiple tests are performed using the same set of data.

Let us begin with a brief refresher on the basics of hypothesis testing.

The first step is to set up the null hypothesis. Conventionally, this expresses a conservative position (e.g. 'there is no change'), in terms of the values of one or more parameters of a population distribution. For instance, in the population of patients with some particular medical condition, the null hypothesis may be: 'mean recovery time (μ_1) after using a new treatment is the same as mean recovery time (μ_2) using the standard treatment'. This is written symbolically as $H_0: \mu_1 = \mu_2$.

Then we specify the alternative (or 'experimental') hypothesis. For instance, 'mean recovery time using the new treatment is different from mean recovery time using the standard treatment'. We write this symbolically as $H_1: \mu_1 \neq \mu_2$. Though we would usually hope that the new treatment generally results in a shorter recovery time ($\mu_1 < \mu_2$), it is conventional, in clinical contexts, to test with a two-sided alternative. We must also specify an appropriate level of significance (usually 0.05, but see the answer to QUESTION 16.1), and a suitable test statistic – for example, the one specified by the two-sample t-test of a difference of means.

Before proceeding, we must confirm the fitness for purpose of the chosen test. The two-sample t-test assumes that the data in each group come from

A Panorama of Statistics: Perspectives, Puzzles and Paradoxes in Statistics, First Edition.
Eric Sowey and Peter Petocz.
© 2017 John Wiley & Sons, Ltd. Published 2017 by John Wiley & Sons, Ltd.
Companion website: www.wiley.com/go/sowey/apanoramaofstatistics

a normal distribution. If this is not a reasonable assumption, we may need to transform the data to (approximate) normality (see QUESTION 14.4).

Next, we collect data on the recovery times of some patients who have had the new treatment, and some who have had the standard treatment. It is important for the validity of conclusions from statistical testing that the data collected are from patients sampled randomly within each of the two groups. In performing the test, we gauge how different the mean recovery times actually are.

On the (null) hypothesis that the mean recovery times after the new and the standard treatments are equal, we calculate the probability (called the 'p-value') of finding a difference of means (in either direction) at least as large as the one that we actually observe in our sample data. If this p-value is large, it's very likely that there is no conflict with the null hypothesis that the two sets of recovery times have the same population means. However, we cannot be certain about this conclusion; it might be that the means really are *not* the same, and that our sample data are quite unusual. If, on the other hand, this p-value is small, we would be inclined to the interpretation that the means are not the same and our null hypothesis is incorrect. Again, we cannot be certain; it might be that the means really *are* the same, and our sample data are quite unusual.

How small is 'small' for the p-value? When the p-value is less than the pre-specified significance level. In that case, when we conclude that there is a difference between the *population* means, we could say equivalently that there is a (statistically) significant difference between the *sample* means. It is important to understand that 'significant', here, means 'unlikely to have arisen by chance, *if the null hypothesis is true*'. It does not *necessarily* mean 'practically important' in the real-world context of the data.

As we indicated earlier, any conclusion we draw from a hypothesis test may be wrong. By choosing to use a 0.05 significance level, we admit a 5% chance of rejecting the null hypothesis when it is, in fact, true. Thus, were we to carry out a particular test multiple times with different data, we could expect to make this 'type I error' one time in every 20, *when the null hypothesis is true*. Mirroring the 'type I error' is the 'type II error', where we fail to reject the null hypothesis when it is the alternative hypothesis that is, in fact, true. In practice, it is rarely possible to fix the chance of making a type II error. In this example, that is because we do not know the actual amount by which μ_1 and μ_2 differ. All we know is that they are not equal. This explains why fixing the chance of making a type I error is the focus of the test procedure, even where the type II error may be the more practically important one to avoid.

These are the salient theoretical aspects of a test of a single statistical hypothesis, as presented in introductory textbooks. The procedure seems polished and easy to implement. However, with a little historical background, we shall see that it is by no means uncontroversial. Nor are the results always straightforward to interpret.

---oOo---

Statistical methods for testing hypotheses were developed in the 1920s and 1930s, initially by Ronald Fisher, and subsequently by Jerzy Neyman and Egon Pearson – the same three statisticians whose pioneering role in the theory of estimation we highlighted in CHAPTER 15. In their work on testing hypotheses, these three introduced the terms *null hypothesis, alternative hypothesis, critical region, statistical significance, power* and *uniformly most powerful test*, which are today familiar to every statistician.

There is an important philosophical difference between the approaches of Fisher, on the one hand, and Neyman and Pearson on the other. At the time, it caused a great deal of polemical controversy and personal acrimony between these proponents.

Fisher developed what he called the theory of *significance testing*, which focuses *exclusively* on what we termed, above, the null hypothesis. The purpose of significance testing, said Fisher, is *to reach a conclusion about* the truth of this hypothesis. To proceed, begin by tentatively assuming it is true. Then, under this assumption, compare (a) the probability of getting the test data (or data more extreme than the test data), with (b) a reference value, chosen at the statistician's discretion. As already mentioned, the former probability is nowadays called the 'p-value' and the latter reference value is called the 'level of significance'. After some reflection, Fisher came to the view that it is quite appropriate for the level of significance to be chosen subjectively, even after the p-value has been calculated.

If the p-value is smaller than the level of significance, the test data are unlikely to have been generated under the stated hypothesis. Accordingly, the hypothesis is rejected. Only then, said Fisher, is there a search for another hypothesis to replace the one that has been rejected.

It is worth noting that Fisher offered no theoretical criteria for judging the merits of the test statistic he put forward in each of the hypothesis testing contexts he studied, in marked contrast to his theoretical work on estimation (see CHAPTER 15). This omission by Fisher was remedied by Neyman and Pearson.

In their alternative to Fisher's approach, labelled *hypothesis testing*, Neyman and Pearson argued that the testing procedure should keep the null and alternative hypotheses simultaneously in view. The purpose of hypothesis testing, they said, is *to make a decision between* these two hypotheses.

They reasoned that a suitable decision procedure could be developed from appraising the relative risks (i.e. probabilities) of both kinds of possible decision errors mentioned above – namely, rejecting the null when it is true (the type I error), and failing to reject the null when the alternative is true (the type II error).

Fixing the risk of the type I error is achieved in the same way as Fisher did, for Fisher's 'level of significance' is nothing but the probability of rejecting a true null hypothesis. However, Neyman and Pearson viewed the choice of level of significance as restricted to a set of standard values (e.g. 0.05, 0.01, 0.001), rather than being open to discretion, as Fisher advocated. Fixing the risk of the type II error is not routinely feasible, since the alternative hypothesis is not specified in exact numerical terms. Nevertheless, that risk can be tabulated for each of a set of parameter values, corresponding to a range of alternative hypotheses. Then, supposing that several competing test statistics are available for the test in question, an optimal selection can be made among them by choosing the one that has – for a given size of type I error – the set with the smallest type II errors.

Biau *et al.* (2010), online at [16.1], contrast the approaches of Fisher and of Neyman and Pearson at greater length. Lehmann (1993) reviews the two approaches in insightful detail and concludes that, *in practice*, they are complementary. For some statisticians, this is sufficient justification to declare the modern textbook account of testing to be a unification of the two approaches. Other statisticians are unconvinced, maintaining that, *in theoretical terms*, the two approaches will always be philosophically incompatible. Thus, they refer – rather negatively – to the modern textbook treatment as a hybrid, rather than a unification, of the two approaches.

---oOo---

It seems to us that when most applied statisticians are at work, they rarely give much thought to the philosophical foundations on which their techniques rest. This lack of attention applies, generally speaking, also to statistics educators.

So, it is a rare statistics course where students learning about hypothesis testing are invited to reflect on questions such as the following. What is the worth of hypothesis tests carried out on non-random samples, such as 'voting' data submitted by readers of online publications? How many missing data values may be tolerated before a standard hypothesis test is no longer worth doing? How common is it to have a uniformly most powerful test? What should be done if such a test is unavailable? What are the essential differences between the Bayesian and the frequentist approaches to hypothesis testing? Why, in testing, is the principal focus of attention the *null* hypothesis – that is, the 'no change, no effect or no difference' situation – which, some argue, is in practice almost never true?

Sometimes, the type II error is more serious than the type I error. This is the case, for instance, in mass screening for cancer, where the type II error of a test on an individual is deciding that the person doesn't have cancer when, in fact, he or she actually does. So, is it ever possible to directly control the size of the type II error in testing?

Knowing how to answer these and similar philosophical questions is very important in developing a deep and secure understanding of statistics and its techniques. With such an understanding, it is easier to recognise and respond to valid criticisms of hypothesis testing.

Here is an example of such a criticism: a null hypothesis (that assigns a specific numerical value to a parameter) can always be rejected with a large enough sample. If, however, this finding represents a (type I) decision error, then this interpretation of the test result will send the investigator off in the wrong direction. That fundamentally limits the usefulness of every technique of hypothesis testing nowadays, since huge samples ('big data') are becoming common in more and more fields (e.g. banking, climatology, cosmology, meteorology, online commerce, telecommunications and analysis of social media). In such situations, it is more fruitful to replace testing by interval estimation, which is as meaningful with 'big data' as with 'little data'.

There are, in fact, further reasons to prefer an interval estimate to a hypothesis test than the likely breakdown of testing in the case of big data. A sustained critique of hypothesis testing has evolved over at least 50 years in the literature of statistics in psychology. This critique has several strands.

Firstly, there is the evidence that test results are too often given erroneous interpretations through faulty understanding of the theory. Here are two examples of common mistakes: the failure to reject the null hypothesis means that the null hypothesis is certainly true; the p-value is the probability that the null hypothesis is false.

Secondly, there is the view that a confidence interval achieves more than a hypothesis test. The reasoning runs like this. Given the endpoints of an appropriate confidence interval, the result of an associated hypothesis test can be deduced (see QUESTION 16.2) but, given the result of a hypothesis test, we cannot deduce the endpoints of the associated confidence interval. Thus, the confidence interval gives more information and, at the same time, it is less liable to misinterpretation.

Krantz (1999) reviews and comments on these critiques and several others. He concludes that, while a few criticisms are misjudged, most are merited, yet he hesitates to recommend the total abandonment of hypothesis testing in favour of interval estimation. More recent writers are not so reluctant. A textbook by Cumming (2012), aimed at psychologists and other scientists who produce meta-analyses, shows how statistical inference can be carried out properly without hypothesis tests. The future implications for statistics education are reviewed in Cumming *et al.* (2002), online at [16.2].

Questions

Question 16.1 (A)

Most statistical hypothesis tests are carried out using a significance level of 5%. But where does this choice of numerical value (almost a statistical icon!) come from?

Question 16.2 (B)

In many contexts, estimation and hypothesis testing can be viewed as two sides of the same coin. Let's explore this proposition. Suppose you have constructed a 95% confidence interval for the mean μ of a population, based on a random sample of data. Now you decide that you would have preferred to use your sample to carry out a test of $\mu = \mu_0$ against a two-sided alternative. How can you use the confidence interval to obtain a result for the test? Could you use this confidence interval if you wanted to carry out a test against a one-sided alternative?

Question 16.3 (B)

Consider the familiar test on the value of the population mean of a normal distribution with known variance, against the two-sided alternative. The power curve for this test has been described as resembling 'an upside-down normal curve'. To what extent is this description correct?

Question 16.4 (B)

If a hypothesis test is carried out using a 5% level of significance against a specific alternative with a power of 90%, and the null hypothesis is rejected, what is the probability that it is actually true?

Question 16.5 (B)

Writing in about 1620 about the game of rolling three dice – at a time when there was as yet little in the way of formal probability theory – Galileo reported that gamblers experienced in this game told him that a total of

9 and 10 can each be obtained in six ways. Yet, it was their perception *over the long run*, that 10 is slightly more likely than 9. (Galileo's resolution of this puzzle is set out in the answer to QUESTION 11.1.)

Let us examine this perception formation over the long run, using modern statistical methods. If the dice are fair, the *theoretical* probabilities of 9 and 10 are 25/216 and 27/216, respectively. In a long run of rolls of three dice, one would expect the *empirical* probabilities of observing a total of 9 or 10 to closely approximate these respective values.

How many rolls of three dice would be necessary to conclude with reasonable confidence that a total of 10 is more likely than 9?

Interpret this question as follows: for a test at the 5% significance level of the null hypothesis that the ratio of chances is 1 : 1, what sample size would give a 90% power of rejecting this, in favour of the alternative that the ratio of chances is 27 : 25?

References

Print

Cumming, G. (2012). *Understanding the New Statistics: Effect Sizes, Confidence Intervals, and Meta-Analysis*. Routledge.

Krantz, D. (1999). The null hypothesis testing controversy in psychology. *Journal of the American Statistical Association* **94**, 1372–1381.

Lehmann, E.L. (1993). The Fisher, Neyman-Pearson theories of testing hypotheses: one theory or two? *Journal of the American Statistical Association* **88**, 1242–1249.

Online

[16.1] Biau, D.J., Jolles, B.M. and Porcher, R. (2010). P value and the theory of hypothesis testing: an explanation for new researchers. *Clinical Orthopaedics and Related Research* **468**(3), 885–892. At http://www. ncbi.nlm.nih.gov/pmc/articles/PMC2816758/

[16.2] Cumming, G., Fidler, F. and Thomason, N. (2002). The statistical re-education of psychology. *Proceedings of the Sixth International Conference on Teaching Statistics* (ICOTS6). At http://www.stat. auckland.ac.nz/~iase/publications/1/6c3_cumm.pdf

17

'Data snooping' and the significance level in multiple testing

It is a fundamental precept of applied statistics that the scheme of analysis is to be planned in advance of looking at the data. This applies to all kinds of procedures. Let's take fitting a statistical model as an example.

The point, in this context, is to ensure as far as possible that the model is vulnerable to rejection by the data. If the data were inspected first, they might suggest a form of model to the investigator, who might then become attached – and even committed – to that model. He or she might then, even subconsciously, twist the data or direct the analysis so that the initially favoured model also comes out best in the end. This kind of subtly biased analysis is especially likely when persuasion (whether social, political or commercial) is the ultimate purpose of the model builder's activity.

To avoid such bias, it is important that the model's form and structure be specified in the greatest detail possible before the data are examined. The data should then be fitted to the model, rather than the model fitted to the data. (For more on fitting a model, see CHAPTER 13.)

What applies to modelling also applies to hypothesis testing: the hypotheses to be tested should be formulated before looking at the data. If the choice of hypothesis (or of statistical analysis, generally) is made after looking at the data, then the process is described as *data snooping*.

In this chapter, we explore the statistical consequences of data snooping in hypothesis testing – in particular, when multiple tests are done using the same set of data. We show why statisticians should be wary of this practice and, yet, why it is almost impossible to avoid.

Data snooping is one of several kindred practices that go by different names in the statistical literature. Selvin and Stuart (1966) distinguish data snooping, data hunting and data fishing, which they refer to collectively as

A Panorama of Statistics: Perspectives, Puzzles and Paradoxes in Statistics, First Edition.
Eric Sowey and Peter Petocz.
© 2017 John Wiley & Sons, Ltd. Published 2017 by John Wiley & Sons, Ltd.
Companion website: www.wiley.com/go/sowey/apanoramaofstatistics

data dredging. Martha Smith's website [17.1], 'Common mistakes in using statistics', has a nice section on data snooping. She writes 'Data snooping can be done professionally and ethically, or misleadingly and unethically, or misleadingly out of ignorance.' We hope to influence you to keep to the first of these alternatives in your statistical work.

We can summarise the textbook procedure for testing a *single* hypothesis test like this. A null hypothesis is set up, expressing a conservative (e.g. 'no change') position – for example, that a particular parameter has the value zero. This is the hypothesis that is to be tested. At the same time, an alternative hypothesis is set up in contrast to the null hypothesis – for example, that the parameter is greater than zero. This is the hypothesis which will be adopted if the null hypothesis is rejected by the test. The null hypothesis is always defined in exact numerical terms, while the alternative is, in general, numerically open-ended.

Evidence is collected in the form of real-world data. If this evidence is unlikely to have arisen if the null hypothesis were true, then the null hypothesis is formally 'rejected' – otherwise, the formal conclusion is 'the evidence is not strong enough to reject the null hypothesis'.

---oOo---

Scientific investigations rarely limit themselves to a single hypothesis. Let's return to our clinical example in CHAPTER 16. Rather than collecting data solely on the recovery times of patients after treatment, we (as medical researchers) will usually gather much more information at the same time: patients' age, sex, body mass index (BMI), blood pressure and pulse rate will be recorded; blood will be taken and cholesterol, glucose and insulin levels measured; and subjective assessments of the patients' state of mind will be obtained via questionnaires. After all, recovery time may depend on many more variables than just the mode of treatment used. Ultimately, we will have a sizable database.

Then, as well as testing whether or not recovery times are different for the two modes of treatment, we may also want to test whether each of the other variables that we have available is related to recovery time – maybe as part of a comprehensive model, maybe as separate tests on individual variables. In this process, we will probably use particular data sets in the database multiple times. Let's say that, in all, we do 25 tests, each with significance level 0.05.

Now we may have a new problem with our testing procedure. Assume, for the sake of illustration, that, in fact, *none* of the measured variables in the database is related to recovery time. Although for each test there is only a

5% chance of a type I error occurring, there is, in the complete set of 25 tests, a higher probability of *at least one* type I error occurring. If the tests were independent (which is not so realistic here), this probability would be $1 - 0.95^{25} = 0.72$. With dependent tests, the probability is likely to be lower, but still well above 0.05. Thus, there is a probability of up to 0.72 of discovering a 'significant' result at least once in the 25 tests. Yet, *any* finding of significance is, by our assumption, illusory.

This illustration shows how multiple tests of hypothesis, performed using a common data set, can inflate the chance of making a type I error to a quite unacceptable level. To counteract this, we could adjust the significance level of each individual test so that the *overall* significance level remains at 0.05. A straightforward way to achieve this is to divide 0.05 by the number of tests to be done (this is known technically as a Bonferroni adjustment – see QUESTION 17.3). Here, $0.05/25 = 0.002$, so an individual test result will be significant if the p-value is less than 0.002.

However, if a very large number of tests is to be carried out, this approach can produce a quite dramatically low value for the significance level of a single test. That will make it very difficult ever to reject the null hypothesis. Thus, Michels and Rosner (1996), writing in *The Lancet* about a situation involving 185 planned tests using the same database, where the overall significance level was to be held to 0.05, say: 'It defies any modicum of commonsense to require a significance level of 0.00027 from a study.'

And this is not the end of the story. Suppose that, after doing our 25 initial tests, we notice that, for female patients over the age of 70, the new treatment seems to work much better than the standard treatment. So, we carry out a further hypothesis test on this subset of our sample, and find a p-value of 0.00015 – surely a significant result, even assessed against the adjusted significance level of 0.002, and one that we could promote among practitioners as evidence for preferring use of the new treatment with older women.

But *is* this latest result, in truth, significant? Well, how many tests will have been done, explicitly *or implicitly*, when we consider our study concluded? Let's see: there are the initial 25 that we carried out earlier, plus this latest one. Now, let's suppose we had picked out females over 70 as one of (say) eight subgroups to test (two sexes and four age groups). True, we didn't actually do the other seven tests, because it was fairly obvious by inspection that there would be no significant result. Also, what about the other combinations of explanatory variables that we reviewed (BMI, blood pressure, higher than usual lipid levels, etc.)? Maybe we should allow for testing these in the subgroups as well. But then, allowing for all these extra tests, the adjusted significance level would get too small, and our result for females over 70 would (unfortunately) no longer be significant. Better, then, to forget about our many exploratory investigations that turned out insignificant, and report only those that are significant. We are much more likely to get such a report published!

Now we *have* slipped over the line into the statistically unethical behaviour that data snooping can represent! As the American Statistical Association says in its *Ethical Guidelines for Statistical Practice*, online at [17.2]: 'Selecting the one "significant" result from a multiplicity of parallel tests poses a grave risk of an incorrect conclusion. Failure to disclose the full extent of tests and their results in such a case would be highly misleading.'

An honest (and professionally defensible) strategy is to do and report only the tests that we specify in advance, adjusting the significance level we use in each test for the number of tests. There is no objection to undertaking the fishing expeditions that tempt us as we proceed with our pre-specified agenda of tests. Indeed, it may be hard to resist their allure. However, the results of fishing expeditions should be reported as such, and the process described, perhaps, as 'hypothesis generation' rather than 'hypothesis testing'.

A valid way to test a hypothesis thrown up in a fishing expedition is to seek out a new set of data. Alternatively, if we initially have a large set of data (meaning many subjects, rather than many variables), we could divide the set randomly into two parts, using one part to generate hypotheses and the other to test them.

In recent years, there has been a movement in many professional fields (notably in medicine) towards evidence-based practice. It ought to be a matter of deep public concern if the unethical pursuit of data snooping were widespread, for it would raise the suspicion that much of the evidence behind evidence-based practice was, in fact, statistically insignificant. Indeed, there are outspoken researchers who claim that this state of affairs is already real, rather than merely speculative. A striking example is given by Ioannidis (2005), online at [17.3], in a publication provocatively titled 'Why most published research findings are false'. Another such example is given by Simmons *et al.* (2011). A web search of either title opens a deluge of supportive scientific commentary.

Questions

Question 17.1 (B)

A classic example of data snooping in the scientific literature appears in an article that examines the rhythms of metabolic activity of a mythical animal. The author started with a set of randomly generated data representing its metabolic activity, and used standard time series techniques to analyse them. What is the animal, who is the author, and what are the conclusions of the study? And what was the point of doing it?

Question 17.2 (B)

In the context of cyber security, the term 'data snooping' has another meaning. What is this, and what relation does it have to statistical data snooping?

Question 17.3 (B)

The Bonferroni adjustment for multiple testing consists of lowering the significance level for each individual test to α/n (where n is the number of tests carried out, explicitly or implicitly) in order to achieve an overall significance level of at most α for the group of tests as a whole. Explain, in the simplest case of just two tests, how the Bonferroni adjustment works when the tests are dependent (as will be the case when they are carried out using the data on exactly the same set of variables)?

Question 17.4 (B)

In the lead up to the Soccer World Cup in 2010, Paul the Octopus displayed his psychic powers by correctly predicting the outcomes of seven final-round games involving Germany, and then the final between Spain and The Netherlands. More detail about his feat is available online at [17.4]. Paul carried out his predictions by choosing between two identical containers of food marked with the flags of the competing countries. If this were set up as a hypothesis testing situation, what would be the null and alternative hypotheses? What is the p-value from the test? To what extent does the result give evidence for Paul's psychic abilities? In what sense is this result connected with data snooping?

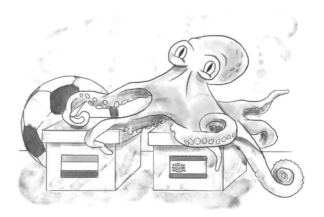

Question 17.5 (C)

In a situation where we carry out a large number of hypothesis tests, there is an unacceptably high chance of finding at least one 'significant' result by chance. Using a Bonferroni adjustment, the overall significance level remains low, but at the expense of requiring very small levels of significance for each individual test. In 1995, two Israeli statisticians put forward a compromise approach to tackling the problem of multiple testing. Who were the statisticians, and what did they propose?

References

Print

Michels, K. and Rosner, B. (1996). Data trawling: to fish or not to fish. *The Lancet* **348**(9035), 1152–53.

Selvin, H. and Stuart, A. (1966). Data dredging procedures in survey analysis. *The American Statistician* **20**(3), 20–23.

Simmons, J.P., Nelson, L.D. and Simonsohn, U. (2011). False-positive psychology: Undisclosed flexibility in data collection and analysis allows presenting anything as significant. *Psychological Science* **22**, 1359–1366.

Online

[17.1] http://www.ma.utexas.edu/users/mks/statmistakes/datasnooping.html

[17.2] http://www.amstat.org/committees/ethics/index.html

[17.3] Ioannidis, J. (2005). Why most published research findings are false. *PLoS Medicine* **2**(8), e124. At http://www.plosmedicine.org/article/info:doi/10.1371/journal.pmed.0020124

[17.4] http://en.wikipedia.org/wiki/Paul_the_Octopus

18

Francis Galton and the birth of regression

It has not been very long since the centenary of the death of one of the founders of our discipline – Francis Galton (1822–1911), student of medicine and mathematics, tropical explorer and geographer, scientist and, above all, statistician. In this chapter, we shall bring to mind something of this truly remarkable man and his statistical contributions. A comprehensive account of Galton's life and work can be had from his own *Memories of My Life* (1908) and from Karl Pearson's three volume *The Life, Letters and Labours of Francis Galton* (1914-24-30). These books, as well as a large collection of Galton's scientific writings, can be read in facsimile at the website [18.1]. A further biography of Galton is cited in CHAPTER 22.

Galton was born into a well-off manufacturing and banking family who were much involved with scientific and literary matters. Members of his immediate family were, in particular, interested in things statistical, his grandfather 'loving to arrange all kinds of data in parallel lines of corresponding lengths, and frequently using colour for distinction', and his father 'eminently statistical by disposition' (*Memories of My Life*, pages 3, 8). His half-cousin was the naturalist Charles Darwin, author of *On the Origin of Species* (1859). Galton showed his high intelligence early. On the day before he turned five years old, he wrote a letter to his sister (quoted in Terman, 1917):

> 'My dear Adèle, I am 4 years old and I can read any English book. I can say all the Latin Substantives and Adjectives and active verbs besides 52 lines of Latin poetry. I can cast up any sum in addition and can multiply by 2, 3, 4, 5, 6, 7, 8, 10. I can also say the pence table. I read French a little and I know the clock.'

A Panorama of Statistics: Perspectives, Puzzles and Paradoxes in Statistics, First Edition.
Eric Sowey and Peter Petocz.
© 2017 John Wiley & Sons, Ltd. Published 2017 by John Wiley & Sons, Ltd.
Companion website: www.wiley.com/go/sowey/apanoramaofstatistics

The story of Galton's early life – particularly his travels and explorations in Eastern Europe and Africa – is a colourful one. After his marriage at age 31 to Louisa Jane Butler, he settled down to a life of scientific studies that included such diverse areas as anthropology, anthropometry, psychology, photography, fingerprint identification, genetics and heredity.

One of his experiments concerned the sizes of seeds, and it turned out to be particularly important statistically, for it led to the birth of the concept of regression. Galton sent several country friends a carefully selected set of sweet pea seeds. Each set contained seven packets of ten equal-sized seeds, with diameters from 15 to 21 hundredths of an inch. Each friend planted the seven packets in separate beds, grew the seeds following instructions, and collected and returned the ripe seeds from the new generation of plants.

Galton first reported the results in an article in *Nature* in 1877, and summarised them in 1886 in his far-reaching paper *Regression towards mediocrity in hereditary stature*. In this paper (page 246), Galton states:

> 'It appeared from these experiments that the offspring did *not* resemble their parent seeds in size, but to be always more mediocre [today we would say 'middling'] than they – to be smaller than the parents, if the parents were large; to be larger than the parents, if the parents were very small. ... The experiments showed further that the mean filial regression towards mediocrity was directly proportional to the parental deviation from it.'

And, in the appendix to the paper (page 259), he writes more specifically:

> 'It will be seen that for each increase of one unit on the part of the parent seed, there is a mean increase of only one-third of a unit in the filial seed; and again that the mean filial seed resembles the parental when the latter is about 15.5 hundredths of an inch in diameter. Taking then 15.5 as the point towards which filial regression points, whatever may be the parental deviation ... from that point, the mean filial deviation will be in the same direction, but only one-third as much.'

Galton then repeated his heredity investigation with human heights. He obtained data on the heights of 930 adult children and their 205 pairs of parents, from family records that he collected by offering prizes. Since women are generally shorter than men, he adjusted the female heights to male equivalents by asking his 'computer' (in those days, a person!) to multiply them by 1.08. Using only large families, with six or more adult children, he tabulated the average height of a child in each family against the mid-parental height (the average of father's and adjusted mother's heights), and found essentially the same results as he had with his seed experiment. He determined that the 'level of mediocrity' (the point where the average height of all children equals the average mid-parental height of all parents) in the population was 68¼ inches, and then defined what he called the 'law of regression' for this context (page 252): '... the height-deviate of the offspring is, on the average, two-thirds of the height-deviate of its mid-parentage.'

Galton had used the term 'regression' for the first time the year before, when he presented these results in person at a meeting of the Anthropological Institute. He had previously used the term 'reversion', but abandoned it because it suggested that the offspring went all the way back to the average of the parents, rather than only part of the way.

FIGURE 18.1, below, is reproduced from the facsimile of Galton's 1886 paper, from which we have been quoting, at www.galton.org. It shows adult child height on the horizontal axis, and mid-parental height on the vertical

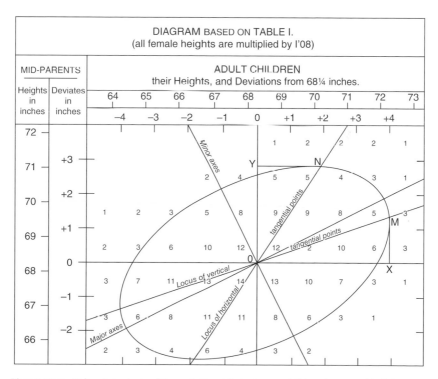

Figure 18.1 Galton's analysis of the relationship between child and parental height. Reproduced with the permission of Gavan Tredoux.

axis (today, following the convention of putting the 'dependent' variable on the vertical axis, we would reverse these axes). The numbers of observations are shown in small digits within the diagram, and the ellipse represents a locus of roughly equal frequencies, in this case connecting the values 3 or 4. About this, Galton wrote (page 254): 'I then noticed ... that lines drawn through entries of the same value formed a series of concentric and similar ellipses.'

In modern terminology, the ellipses represent contours parallel to the base of a three-dimensional bivariate normal distribution. Galton implicitly attributed a normal distribution to the measurement errors in his data. The line through N represents the regression of child height on mid-parental height (see QUESTION 18.5), and the line through M, the regression of mid-parental height on child height. We can see that these two lines are not the same – a point that did not escape Galton.

Initially Galton thought of his discovery of 'regression towards mediocrity' as simply a characteristic of heredity. However, by the time he published his book *Natural Inheritance* in 1889 he understood it for what it really is – a statistical artefact, that is, a change signalled by a fitted regression line that does not necessarily represent a change in the real world. This artefact appears not only in work (such as Galton's) with heredity data – it is quite general in contexts involving repeated measures.

Consider a situation where measurements are made on two occasions (call them 'before' and 'after') on a particular attribute of the same population (or of closely similar populations, such as the heights of parents and of their adult children).We are referring here only to attribute populations that are stable, in the particular sense that the 'before' and 'after' (population) means are equal, and the 'before' and 'after' (population) variances are equal. We suppose, moreover, that all measurements are subject to random (normally distributed) measurement error – that is, they are not perfectly correlated between the two occasions. Then, when these repeated measurements are regressed on one another by the method of least squares, it is easy to show algebraically that the slope of the fitted regression line is always less than one.

It is this property of the regression line that produces the phenomenon of regression towards the mean (to give it its modern name). An introduction to the concept of regression towards the mean by Martin Bland is online at [18.2], a good non-technical account can be found in Freedman, Pisani and Purves (2007), chapter 10, and a particularly interesting historical perspective is given in Stigler (1999), chapter 9.

If unrecognised for what it is, this artefact is likely to lead to false interpretations of regression-based results in experimental studies of a kind that is very common in medicine, epidemiology and psychology. There is now an extensive literature showing how this artefact may be adjusted for or circumvented in such studies. A book-length presentation for non-statisticians is Campbell and Kenny (1999).

Galton's studies of 'regression towards mediocrity' represent the beginnings of what we know today as regression analysis. Galton chose the term 'regression' with great care for the quite specific notion he sought to describe, though this is now mostly unknown or forgotten. Regression was not Galton's only contribution to statistics. Far from it; he made statistical contributions in at least a dozen fields, as well as introducing the fundamental statistical idea of correlation. The past hundred years have seen his huge contribution grow, through the work of countless others – possibly beyond even his wildest imaginings.

Questions

Question 18.1 (B)

A student regresses weight in kilograms on height in inches for a group of adult males. Having recorded the results, he decides that it was silly to mix metric and imperial units, and converts the heights to centimetres (using 1 inch = 2.54 cm). Now he can regress weight in kilograms on height in centimetres. Which of the following results will be the same for the second regression as for the first: the intercept coefficient, the slope coefficient, the value of r^2 (the coefficient of determination)?

Question 18.2 (B)

When the coefficient of determination, r^2, equals 1, all the points in an (X,Y) data scatter lie on the least squares regression line of Y on X. When $r^2 = 0$, the least squares regression line of Y on X is horizontal. Sketch the scatter of (X,Y) data points (1,3), (3,3), (5,3), (7,3), (9,3). For the regression of Y on X based on these data, is r^2 equal to 1 or to 0?

Question 18.3 (A)

As a statistician, Galton often carried out statistical estimation (though the actual term was introduced several decades later by R.A. Fisher). Perhaps his strangest activity was to estimate the bodily measurements of 'Hottentot Ladies' on his expedition to South-West Africa (now Namibia) in 1850–1852. How did he carry out this estimation process?

Question 18.4 (B)

The plot of a novel published in 2000 by an English writer: a postgraduate student decides to give up postmodern literary theory and write, instead, a biography about a famous (though fictional) biographer who left notes on three (real) subjects, identified only as CL, FG and HI. FG is Francis Galton, but who are the other two, who is the author of the book and what is its title?

Question 18.5 (B)

In Galton's diagram (FIGURE 18.1, above), the regression line ON of child height on mid-parental height is defined geometrically (N is the point where the tangent to the ellipse is horizontal). Would calculation using the usual least-squares approach result in the identical regression line? Can you explain why or why not?

References

Print

Campbell, D.T. and Kenny, D.A. (1999). *A Primer on Regression Artifacts.* Guilford Press.

Freedman, D., Pisani, R. and Purves, R. (2007). *Statistics*, 4th edition. Norton.

Galton, F. (1877). Typical laws of heredity II. *Nature* **15**, 512–514.

Galton, F. (1886). Regression towards mediocrity in hereditary stature. *Journal of the Anthropological Institute of Great Britain and Ireland* **15**, 246–263.

Stigler, S.M. (1999). *Statistics on the Table: The History of Statistical Concepts and Methods.* Harvard University Press.

Terman, L. (1917). The intelligence quotient of Francis Galton in childhood. *American Journal of Psychology* **28**, 209–215.

Online

[18.1] http://galton.org (both the Galton papers from the print references are available here)

[18.2] http://www-users.york.ac.uk/~mb55/talks/regmean.htm

19

Experimental design – piercing the veil of random variation

As we highlighted in CHAPTER 2, the study of variation is at the heart of statistics. In almost all fields of mathematics, variation means non-random (i.e. systematic) variation. Statisticians, however, take account not only of non-random variation, but also of random (i.e. chance) variation in the real-world data they work with. This contrast, indeed, distinguishes statistics from mathematics. What's more, the two types of variation that statisticians deal with are almost always present simultaneously. Sometimes, it is the influence of the random variation which is dominant in a particular data set – as, for example, in day-to-day movements in the price of a particular share on the stock exchange. Sometimes, it is the other way round – as, for example, in the monthly value of sales of ice cream in a particular city, where the regular seasonal pattern city-wide dominates random local variation.

It is useful, for what follows, to think of the patternless chance variation as being overlaid, like a veil, on some underlying pattern of systematic variation. A prime goal of statistical analysis is to get behind this veil of random variation in the data, so as to have a clearer picture of the underlying pattern (i.e. the *form*) of systematic variation in the variable or variables of direct interest. This goal is pursued with reference not just to the data at hand, but also (by using appropriate techniques of statistical inference) to the population from which the data came. Where more than one variable is of direct interest, there is an additional motive for getting behind the veil – to identify the degree of stability (i.e. the *strength*) of the pattern of relations between the variables.

It follows that the veil of random variation is actually a kind of obstructive nuisance. In most real-world settings, there is also a second kind of nuisance variation. It is the variation of systematic variables that *are not* of direct

A Panorama of Statistics: Perspectives, Puzzles and Paradoxes in Statistics, First Edition.
Eric Sowey and Peter Petocz.
© 2017 John Wiley & Sons, Ltd. Published 2017 by John Wiley & Sons, Ltd.
Companion website: www.wiley.com/go/sowey/apanoramaofstatistics

interest, but whose influence is nevertheless present. Statisticians aim in various ways to neutralise the impact of both these kinds of 'nuisance variation', so that they can get on with their real objectives – to study the form and strength of the systematic variation in the variables that *are* of direct interest.

---oOo---

All practical statistical studies fall into one of two categories: non-experimental and experimental. Non-experimental studies are sometimes termed observational studies (a rather inexpressive term, since the word 'observation' also turns up in reports on experimental studies!). In an experimental context, the effects of some intervention by the experimenter on a set of experimental units (which may be animate subjects or inanimate objects) are recorded. These data are then analysed to determine whether or not it is likely that the intervention affects the experimental units in some systematic fashion. In a non-experimental context, by contrast, data on variables of interest are collected in the real world, however they occur; there is no intervention. Intuitively, it should be clear that there is greater potential to neutralise nuisance variation successfully when one can (at least partially) control both the source and the intensity of that nuisance variation – which is what a well-designed intervention is intended to do.

We now focus on experimental contexts. Suppose we are interested to know whether a theoretical scale of difficulty that is used to classify particular cases of some task is valid – that is, that the tasks labelled 'easy' are *actually* found by people to be easy, and that those labelled 'difficult' are *actually* found to be difficult. To investigate this, we might take as a null hypothesis that the theoretical scale *is not* valid – that is, tasks labelled 'easy' and 'difficult' are actually perceived in much the same way. Then we are interested to see if the data we collect will reject this hypothesis in favour of the one-sided alternative hypothesis.

Let's take a specific context – Sudoku puzzles – and consider a tentative approach to developing the hypothesis test. If you are unfamiliar with Sudoku puzzles, there are countless websites where you will find them described.

Choose, say, 60 experimental subjects, and give each a Sudoku puzzle to solve, where these puzzles are drawn from a pool containing puzzles labelled 'easy', 'medium', 'hard' or 'diabolical'. For subsequent analysis, we shall proxy the four states of the 'theoretical level of difficulty' by numbers on an ordinal scale. It is common in such contexts to use values in arithmetic progression (e.g. 1, 2, 3, 4, respectively), though it could reasonably be argued that values spaced progressively more widely (e.g. in geometric progression) will more realistically proxy the increasing theoretical level of difficulty of the four kinds of Sudokus. Next, we record how many minutes it takes each subject to complete his or her puzzle.

In this way, we collect a numerical observation for each subject on the variable 'theoretical level of difficulty' and on the variable 'time taken'. If the scatter plot of time taken against (increasing) theoretical level of difficulty for these 60 subjects has a positive slope, it suggests that 'time taken' varies directly with 'difficulty'. We would interpret this result to mean that the theoretical scale of Sudoku difficulty is valid.

If, on the other hand, the scatter plot is roughly horizontal (i.e. has zero slope), this suggests that there is no systematic relation between 'difficulty' and 'time taken'. We would interpret this to mean that the theoretical scale of Sudoku difficulty is not valid.

These interpretations, specific to the sample of 60 subjects involved, could be generalised for the population of all Sudoku solvers by applying a formal significance test to the slope of a line of best fit to the sample scatter plot. If the slope of this line were significantly greater than zero, the null hypothesis ('the theoretical scale of difficulty is not valid') would be rejected.

Any formal significance test mentioned in this experimental context has, as its theoretical foundation, a statistical model of the experiment. It is worth recalling, from CHAPTER 13, that such a model includes both a deterministic component (comprising one or more systematic variables that influence the time taken to solve a Sudoku) and a random component (which we have likened here to an overlaid veil).

The foregoing interpretations would be entirely valid if slope in the scatter plot reflected *solely* an intrinsic population relation between time taken and theoretical level of difficulty. Unfortunately, this proposition is not necessarily true – and not only on account of random variation. Why not? Because, apart from the theoretical level of difficulty (our focus variable), there are also several systematic *nuisance* variables in this setting that have not been taken into account. Here is one such variable: how much prior experience, on a binary scale (more experienced/less experienced), each subject has in solving Sudokus. To see how this categorical variable could influence the test outcome, consider two contrasting scenarios.

If the more experienced solvers all happened to get easy puzzles, and the less experienced solvers all got diabolical ones, the scatter plot of time taken against theoretical level of difficulty would – *already for this reason alone* – have a positive slope.

Now suppose that the more experienced solvers all happened to be assigned diabolical Sudokus, and the less experienced solvers all assigned easy ones. Then it could well be that the two groups take, on average, roughly the same amount of time to solve their puzzles. In that case – and *already for this reason alone* – the slope of the scatter plot might be close to zero, or even negative.

In other words, trend shape in the scatter plot could be the consequence of uncontrolled systematic nuisance variation, rather than a reflection of some intrinsic population relation solely between time taken and level of difficulty.

---oOo---

How might either of these two 'extreme' allocations of puzzles to subjects arise? If the experimenter is the one who does the allocation, there is always a risk of bias (even if it is only unconscious bias), and it is not hard to think of reasons why this might be so. A simple way to counter this risk is to take the allocation out of human hands, and use a computer-generated set of random numbers (see CHAPTER 11) to randomly divide the set of subjects into four groups. A group is then chosen at random from the four, and all members of that group are assigned an easy Sudoku. The next group is randomly selected and assigned a medium Sudoku, and so on.

Now the tentative experimental approach we described initially has been improved. We have created, albeit in a simple way, a *designed* experiment that has neutralised (to a large extent) the influence of the 'prior experience' nuisance variable. How has it been neutralised? By nullifying its systematic influence in the statistical model of the experiment. In informal language, you can think of this as deleting the variable from the deterministic component of the model and adding it into the random component. In the technical language of statistics, the influence of the 'prior experience' variable has been *randomised.*

The resulting improved procedure is called a *completely randomised design with one factor.* Now that the 'prior experience' nuisance variable has been effectively (we trust!) dealt with, the single factor relates to the variable of interest in the intervention used. In this case, the intervention is assigning a Sudoku puzzle to be solved, and the factor is the level of difficulty of the puzzle. This approach can be generalised to two (or more) factors, where each subject does two (or more) different tasks with parallel theoretical scales of difficulty, e.g. a 9×9 Sudoku puzzle and a 6×6 Sudoku puzzle.

You will have noticed that, in order to form the four groups, the randomised design just described needs no knowledge of the actual level of experience that each subject has. If the levels of experience are, in fact, known, then a more efficient design is available (i.e. one which is more likely to lead to rejection of an incorrect null hypothesis). As always in statistical inference, the more correct information that is brought to bear on a problem, the more reliable the inference. The more efficient design is a *randomised block design*.

In the present context of a randomised block design with a single factor, the 'blocks' are two *internally homogeneous* groups of subjects – 'more experienced solvers' and 'less experienced solvers' – which must be set up first. The setting-up process is called *blocking*. By an 'internally homogeneous group of subjects', we mean a group having less variability within the group – in the subjects' puzzle-solving experience – than in the population of all the subjects taken together. By an extension of this definition, if the population is divided into two non-overlapping groups, each of which is internally homogeneous, the variability *within* each group *is likely to be* less than the variability *between* the groups.

Sudokus of all four levels of difficulty are then assigned at random to the subjects within each block. It is the lesser variability within the blocks, with regard to the subjects' puzzle solving experience, relative to the variability between the blocks that gives this design its advantage. In the technical language of statistics, we say that the randomised block design avoids *confounding* the effect of the subjects' puzzle solving experience with the effect of the level of difficulty of the Sudoku puzzle itself.

When there are two or more systematic nuisance variables to neutralise by blocking, a randomised block design can become very complicated, and may require a very large number of subjects for reliability of statistical hypothesis tests. Such a large number of subjects may be prohibitively expensive to seek out. For the case of one factor of interest and two nuisance variables, a more efficient experimental design is available – that is, one which requires fewer subjects than the corresponding randomised block design. It is called the *Latin square design*. For more on this design, showing also how it controls the influence of a nuisance variable, see QUESTION 19.3.

To this point, we have been describing experimental designs for testing a null hypothesis in an experimental context involving a single relationship of direct interest – in our example, the relationship of level of puzzle difficulty to time taken to solve it. However, these same designs can be applied to testing a null hypothesis comparing two relationships, to assess whether they are, or are not, significantly different. Two examples of such contexts are: deciding which of two chemical processes for producing a particular compound provides the best quality product; and deciding whether a new drug is, or is not, more effective than an existing drug for treating a particular illness.

We might now go on to describe the statistical tests that are appropriate to testing hypotheses under each of these experimental designs, and to discuss some of the more elaborate designs that have been devised. However, the technicalities involved would quickly take us beyond the intended purpose of this Overview. A good introductory treatment can be found in chapter 11 of Davies *et al.* (2005). A well-regarded tertiary level textbook, with a bias to engineering applications, is Montgomery (2013).

Questions

Question 19.1 (B)

The pioneering statistical ideas and methods for the design of experiments are due to R.A. Fisher, one of the founders of modern statistical inference. On page 11 of his path-breaking treatise, Fisher (1935), he introduced his subject in this memorable way:

> 'A lady declares that by tasting a cup of tea made with milk, she can discriminate whether the milk or the tea infusion was first added to the cup. We will consider the problem of designing an experiment by means of which this assertion can be tested ... Our experiment consists in mixing eight cups of tea, four in one way and four in the other, and presenting them to the subject for judgment in a random order ... Her task is to divide the 8 cups into two sets of 4, agreeing, if possible, with the treatments received.'

Fisher did not entirely invent this setting – it refers to an actual occurrence. Who was the 'lady tasting tea'? And what were the real circumstances on which Fisher's account is based?

Question 19.2 (B)

Where is the Rothamsted Agricultural Research Station? What part did it have in the development of modern statistical inference prior to 1940 – in particular, by R.A. Fisher?

Question 19.3 (B)

A scientist is interested in studying a particular agricultural relation – how crop yield varies with different amounts of a new chemical fertiliser, measured in grams per square metre. It is not adequate, for this purpose, simply to sow the crop in several plots of ground, then apply differing amounts of fertiliser to each plot, and then measure the weight of crop harvested from each plot. That is because there are inevitably other nuisance variables in the background that also affect the experimental outcome.

When there are two nuisance variables (for instance, the amount of moisture in the soil and the depth at which the seeds are sown), it is statistically efficient to use a Latin square experimental design. What is special about a Latin square design, and what is Latin about it?

Question 19.4 (B)

What is a 'placebo'? In what kinds of experimental contexts is a placebo useful? Are there situations where there is a caveat on the use of a placebo? And, just by the way, what is the linguistic connection between the words 'placebo' and 'caveat'?

Question 19.5 (B)

Name three disciplines from the physical, biological or social sciences where relationships among variables of interest are most commonly examined via experimental studies. Name three scientific disciplines where relationships among variables are most commonly studied non-experimentally. Name three scientific disciplines where experimental studies and non-experimental (also called 'observational') studies are both common. Is anything interesting revealed by this review?

References

Print

Davies, M., Francis, R., Gibson, W. and Goodall, G. (2005). *Statistics 4,*
3rd edition. Hodder Education (in the UK school textbook series
MEI Structured Mathematics).

Fisher, R.A. (1935). *The Design of Experiments*. Oliver and Boyd.

Montgomery, D.C. (2013). *Design and Analysis of Experiments*, 8th edition.
Wiley.

20

In praise of Bayes

It is hard for us today to capture the intensity of the intellectual struggles that past pioneers in any field of knowledge engaged in as, with insight, creativity and sheer hard work, they laid the foundations of that field. However, we can improve our understanding of these struggles if we have some historical knowledge. That is why there are vignettes from the history of statistics in many places in this book.

CHAPTER 22, in particular, gives a broad perspective over some 400 years on the development of statistical inference. In the main, this is a history of frequentism in statistics.

Frequentism is a conceptual framework for statistical theory which takes its name from one of its fundamental axioms – that probability is best defined *objectively* as an empirical *relative frequency*. Unfortunately for any hope of a tidy intellectual evolution of the field, some 18th century statistical thinkers saw scope for an alternative framework for statistical theory, using as a fundamental axiom the *subjective* definition of probability. This conceptual framework has become known as Bayesianism, as we explain below.

Today, frequentism and Bayesianism are thriving as rival paradigms, both for designing theoretical techniques and for interpreting the results of applying those techniques to data. In this chapter, we look at the origins of Bayesianism and show why Bayesian inference is sometimes (its practitioners would say 'always') more appealing than the frequentist alternative.

---oOo---

It all began in 1654, the year that Blaise Pascal sought the aid of his great mathematical contemporary, Pierre de Fermat, to solve at last a

A Panorama of Statistics: Perspectives, Puzzles and Paradoxes in Statistics, First Edition.
Eric Sowey and Peter Petocz.
© 2017 John Wiley & Sons, Ltd. Published 2017 by John Wiley & Sons, Ltd.
Companion website: www.wiley.com/go/sowey/apanoramaofstatistics

fundamental question that had been studied only partly successfully for centuries: given an observed real-world situation (call it the 'cause'), where each of the possible outcomes (call it an 'effect') is a chance event, how can we *systematically* assign a quantitative measure – a probability – to the chance of occurrence of any one of these 'effects' of the observed 'cause'?

During the following century, several alternative approaches to answering this fundamental 'probability problem' emerged. A major obstacle to arriving at a comprehensive general solution was that measurement is an elusive notion in the context of probability. Mathematical principles for assigning probabilities *objectively* were devised. However, the self-evident fact that, in daily life, people commonly make their own *subjective* assessments of probabilities, could not be ignored; yet, there seemed to be no systematic principles that governed the formulation of such subjective probabilities. Worse still, there was no reason why subjective and objective probability assessments of the same event would be consistent with one another. Thus, by 1760, the hoped-for comprehensive solution of the 'probability problem' was still rather in disarray.

At the same time, little headway had been made with another fundamental problem, dubbed the 'inverse probability problem'. The fact that it was easy to state made the seeming intractability of its solution all the more galling to those who struggled with it.

The inverse probability problem can be expressed straightforwardly like this. Given an observed chance outcome (call it the 'effect') of some real-world situation (call it a 'cause'), and knowing the full set of possible real-world situations ('causes') that could have given rise to this outcome ('effect'), how can we *systematically* assign a probability to the chance that the observed 'effect' came from a particular one of the set of real-world 'causes' that could have produced that 'effect'?

If you contrast the relevant wording of the first and fourth paragraphs of this subsection, the reason for the name 'inverse probability' problem should be clear.

In 1763, a remarkable paper on probability was presented at a meeting of the Royal Society in London. It had been written by the Reverend Thomas Bayes (1702–1761), who had earned his living as a church minister in the English town of Tunbridge Wells. In company with many amateur mathematicians and scientists of that era, his research was done in his spare time and was unpaid. Bayes' paper was titled 'An Essay towards solving a Problem in *The Doctrine of Chances*' (he was referring to an early text on probability, *The Doctrine of Chances*, published by Abraham de Moivre in 1718). Bayes' *Essay* was essentially complete (though perhaps not yet polished) at his death, when it came into the hands of his literary executor, Richard Price.

Though Bayes' discussion was difficult to grasp, Price understood that the problem on which Bayes had made progress was, in effect, the inverse probability problem. Price thought this important enough to bring it to the attention of the Royal Society, of which he was a member (as Bayes had been, too). You can see the original version of the essay online at [20.1].

To thinkers who followed Bayes, it seemed that Bayes had implicitly achieved more than to propose a constructive path to solving the inverse probability problem. He had also shown that there was scope, in practice, for synthesising objective and subjective numerical probabilities (so troublesomely distinct as concepts). Bayes' discussion implied that an initial *subjectively-evaluated* probability could be 'revised' in the light of further *objective* probability information from the real world, thus producing a probability assessment that was a meaningful blend of both evaluations.

Many advances in probability theory – and, indeed, in statistical inference – grew out of Bayes' *Essay* over the next 150 years. There is a comprehensive overview in a technical book by Dale (1999). Bayes would be astonished!

Here we shall focus only on a single very important formula in modern probability theory that can be traced back in spirit to Bayes' *Essay*, though it does not actually appear there. This formula – now variously called Bayes' formula, Bayes' rule or Bayes' theorem – is central to solving problems in inverse probability.

All such problems involve conditional probabilities. A conditional probability is the probability that an event A occurs, given that the 'conditioning' event B has occurred. This is written as $P(A|B)$, and is formally defined as $P(AB)/P(B)$ – the ratio of the probabilities of the joint event and the conditioning event (this definition requires that $P(B)$ is not equal to zero). From this, we may write $P(AB) = P(A|B)P(B)$, a useful way of expressing the probability of a joint event.

All problems in inverse probability involve inversion of event and conditioning event. Let's illustrate this inversion with a problem that Bayes himself posed. As a man of the Church, Bayes was interested in the question, 'What is the probability that God exists, given all that I see around me in the extant world?' Bayes realised that P(world exists | God exists) = 1 (since God can make whatever He likes), but what Bayes wanted to know was P(God exists | world exists).

Bayes' formula, which expresses the relation of a conditional probability to the corresponding inverse conditional, can be written in various forms. We shall illustrate one of these in the context of forensic probabilities. A court of law is concerned with whether a suspect is guilty (G) or innocent

(I), given the presence at the scene of a crime of some form of evidence (E), such as a fingerprint, a bloodstain or a DNA sample.

With some reasonable assumptions, we can usually evaluate P(E|G) and also P(E|I), the probability of the evidence being present given the guilt, or the innocence, of the suspect. But what the court actually aims to assess is the inverse of this, P(G|E), the probability that the suspect is guilty given the evidence. To find an appropriate expression, we need a few steps of simple algebra.

We begin with the identity in terms of joint probabilities:

$$P(GE) = P(EG)$$

We can rewrite this equation as

$$P(G|E)P(E) = P(E|G)P(G) \qquad (1)$$

Similarly, since

$$P(IE) = P(EI)$$

we can write

$$P(I|E)P(E) = P(E|I)P(I) \qquad (2)$$

Then we can form the ratio of the left hand sides and the right hand sides of equations (1) and (2), cancelling the (non-zero) term P(E), to show Bayes' formula in the following form:

$$\frac{P(G|E)}{P(I|E)} = \frac{P(E|G)}{P(E|I)} \times \frac{P(G)}{P(I)}$$

In this version, the formula has an interesting theoretical (as well as practical) interpretation. The second term on the right hand side is termed the *prior* odds of guilt – that is, the odds of guilt before any evidence is considered. (You may recall that the odds of an event is the ratio of the probability that the event occurs to the probability that it does not occur. Odds is a measure of chance alternative to the usual 0 to 1 scale of probability.)

The first term on the right hand side is the ratio of the probability that the evidence is present, given that the suspect is guilty, to the corresponding probability, given that he is innocent. It is referred to as the *likelihood ratio* for the presence of the evidence. The term on the left hand side is the odds of being guilty, rather than innocent, given the evidence that has been considered. This is referred to as the *posterior* (or *revised*) odds of guilt.

In summary, Bayes' formula can be expressed as:

$$\text{Posterior odds} = \text{Likelihood ratio} \times \text{Prior odds}.$$

To see how the formula is applied in a legal context, consider this scenario. A suspect is on trial for a murder committed in Australia. A bloodstain found at the murder scene is of type AB−. The victim did not have this blood type, but the suspect does have AB− blood. What can we conclude from this piece of evidence?

First, we can make an assessment of the prior odds of guilt. Suppose there are only 50 people who could conceivably have been responsible for the murder; thus, we shall take the prior odds of guilt as 1/50 or 0.02. Next, we can consider the strength of the evidence. In Australia, only around 1% of people have this rarest type of blood. So $P(E|G) = 1$, while $P(E|I) = 0.01$, and the likelihood ratio is 100, representing a moderate strength of evidence. The posterior odds of guilt is thus $100 \times 0.02 = 2$, indicating that, after taking the evidence into account, the suspect is twice as likely to be guilty as innocent. This is quite an increase on the prior odds!

An important strength of this technique is that it can be applied repeatedly to take account of further *independent* kinds of evidence – for instance, a witness report that the murderer was a man, or the discovery of a fingerprint on the murder weapon. In each step, the current odds of guilt is multiplied by the likelihood ratio of the evidence, to produce a revised posterior odds that the suspect is guilty.

However, the technique evidently cannot proceed without an initial estimate of the probability of guilt. It may be difficult to obtain agreement on such a prior probability. An *objective* 'frequentist' argument could provide a starting point, as in our explanation above. However, for many events, such an initial assessment of chances has to be made in terms of *subjective* probability, because there is simply no other reasonable way to determine their probability. For example, the probability that a particular swimmer will win a gold medal at the next Olympic Games can, in principle, only be assessed subjectively. To many people, including some statisticians, this seems 'unscientific', and so the entire Bayesian procedure for revising odds ratios is dismissed. This seems quite an extreme reaction, given that – whatever element of subjectivity is injected first – that initial element is progressively synthesised with likelihoods evaluated from *multiple* pieces of accrued objective evidence.

In the last century, the notions that prior probabilities can be assessed subjectively and that, more generally, the theory of statistical inference should embrace subjective probabilities were, regrettably, the cause of frostiness and even acrimony among academic statisticians on many occasions. Rather than collaborating intellectually, Bayesians and frequentists took refuge in separate 'camps', each side proclaiming the virtues of their own stance and criticising the other. One of us (PP) recalls, as a young academic, attending a conference at which a speaker announced that he would be presenting a Bayesian analysis of a problem. On hearing this, about half the audience stood up and left the room!

Today, the rift is no longer so wide. As the strengths of Bayesian techniques are more widely understood, closer engagement of the 'camps' in applied statistical work is on the horizon.

Indeed, the Bayesian approach to statistics has had many notable successes. One spectacular example was the inferential approach taken by the team of British cryptologists, led by Alan Turing, that ultimately broke the code of the German Enigma message-enciphering machines during the World War II. Winston Churchill claimed that Turing thus made the biggest single contribution to the Allied victory, and historians have estimated that the work of his team shortened the war by at least two years.

The methods of Turing's team, and many other practical successes of Bayes' formula, are described by Sharon McGrayne (2012) in her book, *The Theory That Would Not Die*. She shows strikingly how a simple rule that, in effect, formalises the notion of learning from experience, has been applied to a vast range of areas of human activity, from rational discussion about the existence of God to more efficient ways of keeping spam out of your mailbox.

Questions

Question 20.1 (A)

Where is Thomas Bayes buried, and what important statistical institution is located nearby?

Question 20.2 (B)

In this chapter's Overview, we considered the following scenario: a suspect is on trial for a murder committed in Australia. A bloodstain found at the murder scene is of type AB–. The victim did not have this blood type but the suspect does have AB– blood.

Knowing this information, the prosecutor says to the jury: 'In Australia, only around 1% of people have type AB– blood. Hence, the chance that the blood came from someone else is very small – only around 1%. So the suspect is fairly certain to be the murderer, with a probability of about 99%.' What is wrong with the prosecutor's argument?

Question 20.3 (B)

In textbooks of advanced probability you will find a section devoted to so-called 'urn problems'. These are problems that involve selecting balls at random from a collection of different numbers of coloured balls in an urn, as a statistical model for certain real-life sampling situations. (An urn is an opaque vase-like container with a narrow top. In its statistical role, it is another one of the physical artefacts that we write about in CHAPTER 25.)

Many urn problems can be instructively solved using Bayes' formula. Here is one example. An urn contains ten balls, each of which is either red or black. One ball is selected at random and found to be red. What is the probability that it was the only red ball in the urn? [You will need to make an assumption about the process by which the urn was initially filled with red and black balls.]

Question 20.4 (B)

Working with his team of code breakers at Bletchley Park in England during World War II, Alan Turing developed the idea of a scale for measuring strength of evidence. What type of scale was this? What was the unit on this scale? And what was the origin of the name Turing coined for this unit of evidence?

Question 20.5 (B)

In the frequentist approach to interval estimation, a confidence interval for a parameter (e.g. the population mean) is constructed using a procedure that captures the true population mean a specified percentage of the time, in repeated sampling. Suppose that a 95% confidence interval for a population mean is found to be (2.5, 3.5). Can we conclude that there is a 95% probability that the population mean is between 2.5 and 3.5?

What is the usual term for the Bayesian analogue of a confidence interval? What differences in interpretation are there between a numerical confidence interval calculated using the frequentist approach and one using the Bayesian approach?

References

Print

Dale, A.I. (1999). *A History of Inverse Probability from Thomas Bayes to Karl Pearson*. Springer.
McGrayne, S. (2012). *The Theory That Would Not Die*. Yale University Press.

Online

[20.1] http://rstl.royalsocietypublishing.org/content/53/370.full.pdf

Part V

Some statistical byways

21

Quality in statistics

It seems to be obvious that statistics is a *strictly* quantitative discipline. However, that is not so, as we shall explain.

Certainly, statistics is a way of arriving at an understanding of the world using techniques for analysing numerical quantities, either measured or counted. 'Numerical detective work' is the way the great US statistician John Tukey described statistical analysis in his renowned book *Exploratory Data Analysis*.

Before the 19th century, statistics was literally 'state-istics', that is, a description of the state (i.e. the nation) – a description which, moreover, focused heavily on qualitative (i.e. non-numerical) analysis. Questions about a country's productivity, wealth and well-being were answered by analyses based on observed characteristics (without necessarily including any measurements), such as its progress in agriculture and industry and its accomplishments in the arts and architecture. An interesting historical essay by de Bruyn (2004) illustrates how this worked in practice.

Do qualitative analyses still have a place in modern statistics? Indeed, they do. Beginners in statistics may form the impression that it concerns itself only with quantitative data and quantitative analyses. However, qualitative data and qualitative analyses are a vital part of statistics, too.

What exactly do the terms 'quantitative' and 'qualitative' mean in this context? Dictionaries usually define these words by referring back to the terms 'quantity' and 'quality'. The Macquarie Dictionary defines 'quantity' as 'an amount or measure', and 'quality' as 'a characteristic, property or attribute'. Statisticians distinguish data on a quantitative variable from data on a qualitative variable by saying that the former are *values*, whereas the latter are *states*.

A Panorama of Statistics: Perspectives, Puzzles and Paradoxes in Statistics, First Edition.
Eric Sowey and Peter Petocz.
© 2017 John Wiley & Sons, Ltd. Published 2017 by John Wiley & Sons, Ltd.
Companion website: www.wiley.com/go/sowey/apanoramaofstatistics

The quantitative/qualitative distinction is a very basic one; there are more elaborate ways of classifying variables. One such classification scheme was devised in 1946 by the US psychologist Stanley Stevens. His scheme puts variables into four classes: categorical (also called 'nominal'), ordinal, interval and ratio.

Categorical data are associated with a fixed set of non-overlapping categories. Examples of a categorical variable are *city of birth* and *marital status*. Ordinal data (as the name suggests) are assigned a place in an ordered scale according to some criterion. Examples of an ordinal variable are *military rank* and a composer's *opus numbers* (that record the order of composition of musical works, without reference to the time elapsed between their dates of publication). Interval data are numerical values that have a precise position on a continuous scale, with an *arbitrary* zero. They are a step up from ordinal data, in that one can say *how much* further along a scale one item is than another. Examples of an interval variable are *longitude* and *temperature in degrees Celsius*. Finally, ratio data are numerical values that have a precise position on a continuous scale with an *absolute* zero. Examples of ratio variables are *length* and *weight*.

It should be clear from these definitions that interval and ratio variables are quantitative variables. Further, a categorical variable is clearly a qualitative variable. But what can we say about an ordinal variable? Is it quantitative or qualitative? This is a perplexing question, for some ordinal data appear to be quantitative (opus numbers, in our example), while others seem to be qualitative (e.g. military rank).

This issue has caused a great deal of controversy in statistics, especially in regard to psychological data. Psychologists routinely collect ordinal data in their experiments, and are accustomed to assigning numerical ranks to their observations before analysing them. Think, for instance, about the following behavioural question and its numerically ranked responses: Do you smoke? (often 1, sometimes 2, rarely 3, never 4).

These rank data *look like* interval data, but not all statistical calculations (e.g. the arithmetic mean) that are valid with interval data are meaningful with rank data. Why? Because a rank coding of responses is an essentially arbitrary choice. After all, if rarely or never smoking were regarded as personally exceptionally beneficial, then the four responses might, for instance, be coded 1, 2, 4, 8. The subtleties of accommodating ordinal variables in statistical analyses are explained in more detail in a (fairly technical) paper by Velleman and Wilkinson (1993).

Let's look now at some qualitative aspects of modern statistical work that go beyond simply including qualitative variables in analyses.

There are, for example, qualitative issues in *defining* a qualitative variable – that is, defining the 'states' (or 'categories') of the qualitative variable.

In many practical contexts, this can be complicated, and even controversial. For example, in Australia, a person is officially defined as 'employed' if he or she performed at least one hour of paid work in the week prior to the official employment survey. Such a definition has an obvious political implication. It enables the government to report a higher total employment than would be the case if a more stringent definition were adopted. In the world of sport, we distinguish amateurs from professionals. The categorisation of an athlete as 'amateur' was (until the 1980s) an indispensable requirement for participation in the Olympic Games. Not surprisingly, the definition of 'amateur' in this context became a matter of the sharpest dispute.

A remarkable book by Bowker and Star (2000) shows how the definition of categories plays an important role in the outcomes of statistical investigations, with some striking examples involving medical and racial classification.

There are also qualitative issues in including a *quantitative* variable in a statistical analysis, for we need first to decide exactly what to measure and how to measure it.

For example, if we are carrying out a study comparing the effectiveness of two different approaches to learning statistics – a traditional classroom course and an online course – we might initially think of basing conclusions on students' final examination results. However, we know that it is not straightforward to measure the outcomes of learning in this way. This leads us to a consideration of other variables or combinations of variables that might do better. We may, for instance, choose to compare students' attitudes towards statistics at the beginning and at the end of their studies, using an instrument such as the *Survey of Attitudes Towards Statistics*, developed by Candace Schau (online at [21.1]). Again, in a medical context, if we wish to compare two treatments for brain tumours, we might measure survival times, or we might put more emphasis on the quality of life during the survival time, and assess this using a quality-of-life survey – see Carr *et al.* (eds) (2002).

In both of these situations, you will notice that the alternative assessments proposed will produce ordinal data. As already noted above, we would need to be watchful that the statistical analyses applied to these data were practically meaningful.

Sometimes it happens that an investigator finds it too challenging to choose a measure for some particular real-world variable, and so decides simply not to measure it at all. Then, regrettably, the influence of that variable may just be ignored. Consider cost-benefit analysis (already mentioned in CHAPTER 9). Faced with a proposal to 'develop' some land by harvesting the trees growing on it, and then building houses on the cleared land, it may be difficult for a project-assessment authority to measure the benefits of continuing to have trees growing in that particular location – benefits in terms of, say, their prevention of soil erosion, or their appeal as parkland. It is all too tempting to make the decision to approve or to disallow the proposal on the basis of only those economic variables that can be measured easily, such as the costs of harvesting the trees and of building the dwellings, and then weighing these costs against the benefit, evaluated solely as the amount that can be earned from sale of the timber and the dwellings.

As the Nobel Prize-winning economist Joseph Stiglitz has written: 'What we measure affects what we do. If we have the wrong metrics, we will strive for the wrong things.'

Questions

Question 21.1 (A)

Language text (which is qualitative information) can be analysed using frequency counts of letters, words or phrases (that is, in a quantitative way) to attempt to resolve such matters as authorship disputes. Statisticians who contribute in the field of English textual analysis soon learn the order of letters by their frequency of occurrence in English prose. The first 12 letters of this ordered set have been used as a phrase in a variety of contexts. What is this phrase? Can you give a context in which it has been used?

Question 21.2 (A)

FIGURE 21.1 is part of an historic map of London from the late 19th century showing by different shadings (originally, colourings) the socio-economic status of each household. Who created the map? How were the data collected? What current statistical marketing technique is its direct descendent?

Question 21.3 (B)

We have seen the perplexing position of ordinal variables, lying between the quantitative and the qualitative in statistical analyses. Let's examine this further. Suppose two groups of people – A and B – are suffering from the same illness. Those in group A receive treatment T_1, and those in group B receive treatment T_2. Afterwards, each person is asked to respond on a five-point scale – strongly disagree, disagree, undecided, agree, strongly agree – to the statement 'the treatment I received was completely effective'. These responses can be numerically coded as 1, 2, 3, 4, 5. We can use these response data to test whether people who receive T_1 have the same perception of the effectiveness of their treatment as those who receive T_2.

Figure 21.1 A map of central London (extract). Reproduced with the permission of David Thomas.

a) If this hypothesis is tested using a chi-squared test of independence, what assumption is being made about the nature of the response variable? What if an independent-samples *t*-test is used? What test would be more appropriate than either of these?

b) Do any of the tests in part (a) throw light on whether the two treatments are equally effective?

Question 21.4 (B)

In 1973, *The American Statistician* published a paper on sampling with this intriguing title: 'How to get the answer without being sure you've asked the question'. What is the name for the type of sampling that the authors were describing, and in what situations might this type of sampling be useful?

Question 21.5 (B)

Statisticians have made many contributions in military settings. A famous example is an investigation, during World War II, of the survivability of military aircraft hit by enemy fire. Which eminent statistician estimated the probabilities of an aircraft surviving a single hit on different parts of its body? What aspect of the damage data did he particularly notice, and what insightful contribution did that lead to for improving aircraft survivability under fire?

References

Print

Bowker, G. and Star, S. (2000). *Sorting Things Out: Classification And Its Consequences*. MIT Press.

Carr, A., Higginson, I. and Robinson, P. (eds, 2002). *Quality of Life*. Wiley.

de Bruyn, F. (2004). From Georgic poetry to statistics and graphs. *The Yale Journal of Criticism* **17**, 107–139.

Stiglitz, J. (2009). Towards a better measure of well-being. *Financial Times*, London, 13 September.

Tukey, J.W. (1977). *Exploratory Data Analysis*. Addison-Wesley.

Velleman, P.F. and Wilkinson, L. (1993). Nominal, ordinal, interval, and ratio typologies are misleading. *The American Statistician* **47**, 65–72.

Online

[21.1] http://www.evaluationandstatistics.com/view.html

22

History of ideas: statistical personalities and the personalities of statisticians

How much do you know about the historical development of today's theory and practice of statistics?

The modern field of statistics is the cumulative intellectual achievement of hundreds of gifted thinkers over at least the past 400 years and, particularly, since about 1860. To learn about the history of ideas in statistics is to discover the names of those gifted statistical personalities. The scholarly literature of statistics may concentrate on the ideas and give the names only passing regard, but we should not take this as a signal that the names are unimportant. The names *are* important – not in themselves but, rather, for *who they were*, these energetic and creative builders of modern statistics. Knowing something of the personalities of these statisticians, we can hope for insights on 'how they did it'.

In this hope, we statisticians are certainly not unique. It has long been popular to seek, in the personalities and life-paths of the gifted, clues to their remarkable achievements – whether the gifted are thinkers (philosophers, historians, scientists, trainers, etc.) or doers (political leaders, explorers, engineers, athletes, etc.).

Sometimes, this pursuit is disappointing. The lives of the composer Mozart and the painter Rembrandt, for instance, offer few insights on how they created the works of genius that we treasure today. However, there are many other historical personalities whose lives convey much about the sparks that ignited their great achievements. Our insights come principally from two kinds of sources: their own informal writings and revelatory exchanges with their intellectual peers (e.g. personal diaries and private letters); and public documents (e.g. reports of debates and controversies, and biographical essays).

A Panorama of Statistics: Perspectives, Puzzles and Paradoxes in Statistics, First Edition.
Eric Sowey and Peter Petocz.
© 2017 John Wiley & Sons, Ltd. Published 2017 by John Wiley & Sons, Ltd.
Companion website: www.wiley.com/go/sowey/apanoramaofstatistics

All of these sources are, unfortunately, far removed from the settings in which statistics is studied and practised today. Unless steered to these sources by a teacher or reference book, or propelled in their direction by incidental curiosity, few statisticians come upon them. The history of statistical ideas remains a little-visited byway.

Does it matter? You be the judge!

The history of statistical ideas since about 1600 is a grand saga of intellectual endeavour. It tells of achievements in these major areas: how to conceptualise, measure and analyse chance in human experience; how to detect authentic 'big picture' meanings in detailed real-world (and, therefore, chance-laden) data; and – from the knowledge gained in those inquiries – how to evolve a set of inductive principles for generalising the detected meanings, as reliably as possible, to wider contexts.

Without a historical perspective, one has little idea which concepts and principles are recent and which long-established, or which were easily established and which were, for a long time, intractable. Many statistics textbooks so neglect a historical perspective that it could well appear to beginning students that the *entire* body of theory was conceived just recently, and delivered soon afterwards – perhaps by a stork?

In fact, the evolution of modern statistics has been a slow journey with deeply human dimensions.

Three aspects of this journey are worth your attention. First, when, and in what practical circumstances, pivotal ideas were born; second, how challenging it often was for the pioneers of theory just to frame clearly the questions that they wanted to answer; and third, how much rethinking was called for before satisfactory solutions were arrived at. Once you have some perspective over these matters – and, especially, the last one – you will, we feel sure, be cheered by just how much easier is the path to the same knowledge today.

FIGURE 22.1 gives a schematic view of some of the major stages in the laying of the dual foundations on which modern statistical inference rests, together with the names of the scholars with whom the important progressive ideas are associated. These foundations are *probability theory*, and *methods of statistical data summarisation and display*. We should emphasise that the contributions of all those mentioned in this table are far more extensive than what is shown. Only contributions relevant to the themes of the table are included here.

FIGURE 22.2 shows how the structure of statistical inference was erected, after 1860, on the dual foundations in FIGURE 22.1.

To enrich the story traced out in FIGURE 22.2, you may like to browse some of the following references. For each scholar mentioned, there are two items. The first summarises (without too much technicality) his contributions to statistical inference; the second offers insights into his personality and life-path. We have selected these references from a profusion of material, much of it dating from the last 25 years, a period which has seen a wealth of new research in the history of statistics.

Galton	1. Forrest (1974)	2. Galton (1908), online at [22.1]
Pearson	1. Magnello (2009)	2. Porter (2004)
Gosset	1. Plackett (ed, 1990)	2. McMullen (1939)
Fisher	1. Zabell (2001)	2. Box (1978)
Neyman &	1. Lehmann (1993)	2a. Reid (1982); 2b. O'Connor and
Pearson		Robertson, (2003), online at [22.2]
Savage	1. Lindley (1980), online at [22.3]	2. O'Connor and Robertson (2010), online at [22.4]
Tukey	1. Brillinger (2002), online at [22.5]	2. Anscombe (2003), online at [22.6]

Once you've caught the 'bug' on the history of ideas in statistics, where can you turn to go on exploring these ideas more generally? That depends, of course, on where you are currently in your knowledge of the discipline. Here are some suggestions.

If you are currently involved in undergraduate studies, you'll find the book by Stigler (1986) particularly readable on the history of probability theory in the 18th and 19th centuries (the ideas of Bernoulli, de Moivre, Gauss and Laplace), the birth of the normal distribution (the work of de Moivre and Gauss), and the creation of correlation and regression theory (by Edgeworth, Galton and Karl Pearson).

Themes from the evolution of probability theory			Developments in statistical data summarisation and display		
Prior to 1600, and going back in time as far as the Greek philosophers, inquiry into the nature of chance focused primarily on uncertainty in the real world. This was the case, for instance, in assessing the correctness of scientific theories, the success of risky commercial ventures, and the trustworthiness of evidence given in a court of law. No formal principles for quantifying uncertainty developed from these inquiries. After 1600, a new focus emerged. It followed from popular demands to quantify the chance of winning in various kinds of gambling games. This new focus shifted thinkers' emphasis away from anchoring philosophical foundations, to devising rules for assigning a probability to a chance event and for combining probabilities. Yet, it left the foundations unsatisfactorily vague.			Throughout recorded history, those in power have collected data (both numerical and non-numerical) of importance *to them as rulers* – that is, in relation to such activities as military affairs, ownership of land, taxation, production and trade. There was little thought, until the late 1600s, of collecting data for advancing the welfare of the ruled classes. Also, beyond basic tabular presentation, little attention was paid until the late 1700s to systematising data description and summarisation, whether in tabular or graphical form. In the mid-18th century, the established craft of using data to aid political administration of the State became known as 'Statistics'. It was not until the 19th century that statistics was commonly understood as the discipline we know today.		
1600–1760			**1600–1760**		
Galileo Galilei (1564–1642)	Solved, by first principles reasoning, a probability problem on the roll of three dice (1620? – the date is uncertain).	* (a)	John Graunt (1620–1674) William Petty (1623–1687)	Compiled the earliest systematic data for advancing citizen welfare. Graunt tabulated English public health and mortality statistics (1662) and Petty population statistics (1676).	* (e)
Blaise Pascal (1623–1662) Pierre de Fermat (1601–1665)	Pioneers of modern probability theory, as systematic analysts of games of chance (1654–1660). Proposed the 'a priori' definition of the probability of an event.				
Jacob Bernoulli (1655–1705)	The first to use 'probability' as a routine term in the measurement of chance. Adopted the 'a posteriori' (or relative frequency) definition of the probability of an event. Discussed a limited version of the law of large numbers (1713, published posthumously).				
Abraham de Moivre (1667–1754)	Found that the normal distribution approximates well the limit of the binomial distribution for very many trials (1733). Invented the term 'modulus' for a dispersion parameter in the equation of the normal (1738).	* (b)	Johann Süssmilch (1707–1767)	Compiled historical statistics on births, deaths and marriages for cities and towns all over Europe, going back 80–100 years, and tabulated them at length. These form the first extensive array of demographic statistics (1741).	

Figure 22.1 History of ideas I.

1760–1860			1760–1860		
Thomas Bayes (1702–1761)	Published the first attempt to resolve the 'inverse probability problem'. His idea was the precursor of the subjectivist paradigm of statistical inference, an alternative to Bernoulli's frequentist approach (1763, published posthumously).	* (c)	William Playfair (1759–1823)	The pioneer of modern statistical graphics. Devised bar, column and pie charts, and applied the Cartesian coordinate system of pure mathematics to plotting line graphs of bivariate statistical data (1786–1801).	
Carl Friedrich Gauss (1777–1855)	Invented 'least squares' as a way of fitting a mathematical function to data (around 1795). Identified the normal distribution as a good probability model for the distribution of random errors of measurement (1809). Explored the use of the normal model in many real-world data contexts.	* (b)	Adolphe Quetelet (1796–1874)	Attached great significance to the normal distribution as expressing (what he thought of as) a universal systematic pattern in the population spread of each human physical attribute and behaviour. Regarded the mean as signifying 'perfection' in whichever attribute was being graphed (1835).	
Pierre Simon de Laplace (1749–1827)	Developed rules for calculating probabilities of compound events according to the frequentist paradigm. Proved the Central Limit Theorem (CLT) for the case of samples from the binomial distribution. Studied the properties of the normal distribution in detail (1812).	* (b)	John Snow (1813–1858) Florence Nightingale (1820–1910)	Both Snow (a doctor) and Nightingale (a hospital reformer) used graphical displays to increase the impact of their statistical evidence of previously unrecognised sources of disease. Snow, in 1855, plotted cholera cases in Soho on a map, to highlight the likely role of a contaminated water supply. Nightingale, in 1857, plotted polar area charts to show the benefits of clinical antisepsis on healing war wounds.	* (f)
1860–1960			1860–1960		
Alexsandr Lyapunov (1857–1918) Jarl Lindeberg (1876–1932) Paul Lévy (1886–1971) William Feller (1906–1970)	Proved the CLT under ever more general conditions. Independent, identically distributed variables: Lyapunov (1901) and Lindeberg (1922). Independent, non-identically distributed variables: Lévy-Feller (1935). All proofs require a finite population variance. Lévy also identified the family of stable distributions (1923).	* (d)	Etienne Laspeyres (1834–1913) Hermann Paasche (1851–1925)	Designed consumer price index formulae for averaging price changes through time: Laspeyres index with base period weights (1871), Paasche index with current period weights (1874). Both indexes are now widely used.	* (g)

Figure 22.1 (Cont'd)

| Andrei Kolmogorov (1903–1987) | Emphasised probability theory as a branch of pure mathematics, by defining probability in terms of abstract axioms (1933). In this way, he sought to overcome philosophical objections to the 'a priori' and 'a posteriori' definitions, both of which rest on viewing probability as an empirical notion. Also, explored under what most general conditions the CLT is valid (1954). | | | Francis Ysidro Edgeworth (1845–1926) | Slightly redefined de Moivre's 'modulus' as his favoured measure of spread. Conceived the basic idea of the analysis of variance (1885). Advanced the 'stochastic approach' to index number theory (1887). Tentatively applied simple inferential methods to social science data. | * (h) |

Footnotes
(a) See QUESTIONS 11.1 and 16.5.
(b) See CHAPTER 14.
(c) See CHAPTER 20.
(d) See CHAPTER 24.

(e) See QUESTION 12.3.
(f) See QUESTIONS 1.5 and 22.2.
(g) See CHAPTER 7.
(h) On Edgeworth's modulus, see QUESTION 6.1; on the Marshall-Edgeworth price index, see QUESTION 7.5.

Figure 22.1 (Cont'd)

For earlier ideas on chance and probability (from Galileo, Pascal, Fermat and Bernoulli in the 16th and 17th centuries), the engagingly written book by David (1962) can be thoroughly recommended. The contributions of some 20th century statistical pioneers are reviewed in lively fashion by Salsburg (2001).

Short but very informative biographies of most of the 19th and 20th century statisticians mentioned in FIGURES 22.1 and 22.2, together with those of a couple of dozen other statistical pioneers, can be found in the *MacTutor History of Mathematics* online archive (see the index of names at [22.7]).

If you are a postgraduate student in statistics, or in a field where statistical ideas are important, there are many gems to be found in the following three books. Hacking (1990) shows how the evolution of ideas about probability advanced 19th century European society and culture. Stigler (1999) is a collection of 22 stimulating, and sometimes quirky, essays with settings ranging from the 17th to the 20th century.

Also very valuable is Weisberg (2014), which (unusually in this field) is written by an applied statistician. It presents a non-mathematical perspective over the history of probability and statistics, from Pascal's beginnings to the situation today. What makes the book especially interesting is that it elicits, from this history, challenges for tomorrow. These include how to repair a growing gap between the activities, in their increasingly separated worlds, of academic researchers and statistical practitioners in government and business. Academics push forward the frontiers of statistical theory

Themes from the evolution of statistical inference

In the 75 years after 1860, it was gradually realised that enough was understood about probability and about empirical data distributions (from work in progress since 1600) to attempt a grand synthesis of these two threads of knowledge. The goal was to construct, for the first time, a set of formal principles of data-based inductive inference (these principles would parallel the formal principles of deductive inference, which had been substantially worked out by Aristotle and his successors some 2000 years before). Progress with the synthesis was slow at first, given the maze of needed preliminaries – theorems that were easier to state than to prove, and conceptual gaps that were easier to identify than to fill. Since about 1935, progress on all fronts has been rapid. The theory of statistical inference in a frequentist probability framework and in a Bayesian probability framework is essentially complete (for testing the fit of data to pre-specified models), as are the principles of exploratory data analysis (for designing models that fit the data). Nevertheless, controversial issues still abound.

1860–1935

Francis Galton (1822–1911)	Pioneered statistical methods in his extensive studies of heredity, genetics, anthropology and psychology. Introduced the analysis of bivariate statistical relations. Coined the terms: percentile rank, ogive (1875), interquartile range, correlation, regression to the mean (1886), and showed their empirical uses. Using the quincunx (1885), sought to build a statistical model for (his cousin) Charles Darwin's notion of 'natural selection'.	* (a)
Karl Pearson (1857–1936)	A pioneer of biometry and anthropometry. Coined the name 'standard deviation' for the now universal spread measure (1894). Devised a measure of skewness and of bivariate correlation (1896), and a repertoire of models for empirical frequency distributions. Made many advances in statistical inference, including the chi-squared test (1900).	* (b)
William Sealy Gosset (1876–1937)	Whereas Karl Pearson developed inferential theory for *large* samples, his advisee Gosset wanted to formalise inference about the mean from *small* samples, which were standard in his work as an industrial chemist. Gosset (pen name 'Student') partially solved (1908) this problem. R.A. Fisher completed the task (1923).	* (c)

1935–2000

Ronald Aylmer Fisher (1890–1962)	Synthesised and reformulated the work of his predecessors on estimation – for example, on the method of maximum likelihood. However, it is his own ideas (which he defended vigorously against any criticism) that underlie most of the modern principles of estimation, especially for small samples. He created the needed theory as required by his analyses of agricultural and genetic data. His contributions also include the analysis of variance, experimental design, and discriminant analysis. In addition, he proposed systematic methods and protocols for significance tests.	* (d)
Jerzy Neyman (1894–1981) Egon Pearson (1895–1980)	Collaborated (1926 to 1938) on providing a philosophical basis for the theory of hypothesis testing, different from Fisher's. This included attaching importance (denied by Fisher) to the choice of alternative hypothesis, and to the power function for comparing alternative tests of the same null hypothesis. Neyman also developed the theory of interval estimation. Pearson was a prominent historian of statistics.	* (e)

Figure 22.2 History of ideas II.

Leonard Jimmie Savage (1917–1971)	A major early contributor in advancing the then novel notion of subjective probability as the basis of a theory of statistical inference, subsequently called Bayesian inference. This was a new paradigm, contrasting with the work of Edgeworth, Galton, Neyman and Pearson, who built their theory of inference on the frequentist definition of objective probability. Fisher had conceived something (fiducial inference) parallel to this novel approach, but did not succeed in progressing it.	* (f)
John Tukey (1915–2000)	Like Galton and Fisher, a polymath. Tukey's fields were chemistry, physics, mathematics and statistics. His many roles in public policy conditioned him to seek out always-practical analytical tools – that is, tools resilient to invalidity of their restrictive theoretical assumptions. Thus, he invented or promoted many new methods for vivid graphics (e.g. the box plot) and robust estimation (e.g. the jackknife). He recognised (1962) that statistical theory since 1930 had come far on *confirmatory* data analysis – that is, on formally testing statistical hypotheses to see if they are supported by the data – but said very little on how those hypotheses were devised in the first place. This prompted his extensive work on *exploratory* data analysis, a field which he created almost single-handedly over the next 20 years, and which now underpins the modern field of large-scale data mining.	* (g)

Footnotes

(a) See CHAPTER 18 and QUESTIONS 9.3, 14.3, 18.3, 22.4 and 25.3.

(b) See QUESTION 9.5.

(c) See QUESTION 22.5.

(d) See CHAPTER 15 and 16 and QUESTIONS 4.4, 16.1, 19.1, 19.2, 19.3 and 22.5.

(e) See CHAPTERS 15 and 16.

(f) See CHAPTER 20.

(g) See CHAPTER 1 and QUESTION 22.3

Figure 22.2 (Cont'd)

using established quantitative concepts of probability. Practitioners, however, confront very diverse situations, in which the element of uncertainty often has qualitative dimensions not fully captured by formalised probability theory, making the application of academic statistical techniques a fraught matter. The author gives evidence for his views from contemporary statistical practice (e.g. in education, pharmacology, medicine and business). He also urges theorists and practitioners to re-engage, holding up as a model R.A. Fisher's fruitful melding of his contributions to theory and practice (see also the answer to QUESTION 19.2).

By the way, it can be very rewarding to dip into the original works of the statistical pioneers. English translations of most works by the French and German pioneers are available, if reading Latin, French or German is not among your skills. You may be surprised, for example, how directly and informally Galton and Karl Pearson share their thinking with the reader, even while they are still feeling their way towards their eventual technical achievements.

Finally, for the committed enthusiast about the history of statistical ideas, we recommend exploring Peter Lee's extensive website, *Materials for the History of Statistics*, online at [22.8]. Among the diverse links brought together on this website, there are all kinds of unusual things to be discovered. Cited on this website, but meriting separate mention for its rich detail, is John Aldrich's website, *Figures from the History of Probability and Statistics*, online at [22.9]. The contributions of even more statisticians of the past, worldwide, can be found in Heyde and Seneta (eds, 2001). Among the editors and compilers just mentioned, John Aldrich and Eugene Seneta have made their own extensive, and always engaging, scholarly contributions on the history of probability and statistics. Further recent writers who cover this field broadly and whose works repay seeking out are Lorraine Daston, Gerd Gigerenzer, Anders Hald, Robin Plackett and Oscar Sheynin.

When it comes to appreciating the innovative ideas of contemporary statisticians, the best source is often the perspective of the innovator him- or herself. Such perspectives can come to light nicely in informal conversations with the innovators, which are recorded and then transcribed for publication. Since the mid-1980s, published 'Conversations' with leading contemporary statisticians have brought some remarkable ideas and personalities to life on the printed (or digitised) page. The journal *Statistical Science* has included a Conversation in most issues – you can search past issues online at [22.10]. There are (shorter) Conversations also in many issues of the non-technical magazine *Significance*, published jointly by the Royal Statistical Society and the American Statistical Association.

Questions

Question 22.1 (A)

Some statisticians have unexpected hobbies and interests. For example, the eminent British statistician, Maurice Kendall (1907–1983) applied his literary enthusiasm to writing an experimental-design pastiche of Henry Wadsworth Longfellow's poem, *Hiawatha* (reprinted as Kendall, 2003). W. Edwards Deming (1900–1993), the US statistician who promoted statistical quality control internationally, composed church music. And Persi Diaconis, Professor of Statistics at Stanford University, USA, is an expert conjuror.

a) Maurice Kendall is associated with another literary contribution of some repute, this time as ghost writer. It contains a now quite famous statement about the nature of statistics, to the effect that it is not the numbers that matter but, rather, what you do with them. Where did this statement first appear, and who was the publicly credited author?

b) Which US statistician, active during the 20th century, had typography as a hobby, and how did he apply that hobby to celebrating the importance of the normal distribution to scientific observation and experimentation?

Question 22.2 (A)

What is the historical context of the diagram in FIGURE 22.3? Who constructed the original version of the diagram, and for what purpose? What is the name of this type of diagram?

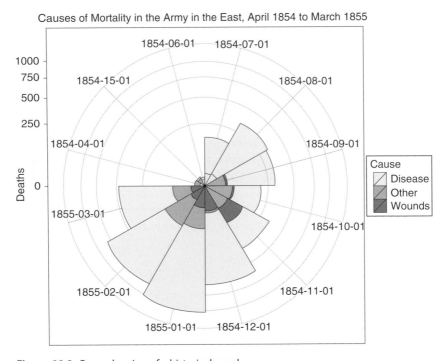

Figure 22.3 Our redrawing of a historical graph.

Question 22.3 (A)

a) An arithmetic mean combines all the numerical values of the data in calculating the average, while finding a median requires only that the values be arranged in order of size. But what type of average involves first ordering the values and then combining some of them? In what situation would such an average be useful?

b) In 1972, John Tukey was a co-author of a major study of different sample estimators of the central value of a symmetric population distribution.

What was the title of this study, what was its objective, and what did it have to say about the average asked about in part (a)?

Question 22.4 (B)

The scene is an English country livestock fair about a hundred years ago. A large animal is displayed, and there is a competition to guess its weight when it has been slaughtered and dressed (i.e. prepared for cooking). For a small sum, anyone can submit a guess and compete for the prize for the most accurate guess. Later, a famous statistician examines the recorded guesses and writes a short article based on them. Who was the statistician, what type of animal was the centrepiece of the competition, and what was the statistician's conclusion?

Question 22.5 (B)

a) The man who called himself 'Student' in almost all of his scholarly publications was William Sealy Gosset. What was Gosset's day job when he *partly* solved the problem of finding the exact probability density function of what we know as Student's t-distribution? Why did he publish his result (Student, 1908) under a pseudonym?

b) But should it really be *Student's* t-distribution? Here is some historical perspective to clarify this question:

Given a random variable, X, normally distributed as $N(\mu, \sigma^2)$, with σ^2 unknown, and given the mean, $\overline{X}$, of a random sample of size n from this population, we know the t-statistic for testing the null hypothesis H_0: $\mu = 0$ as $t = (\overline{X} - \mu)/(S/\sqrt{n})$, where $S^2 = \Sigma(X - \overline{X})^2/(n-1)$. In his 1908 paper, however, Student (i.e. Gosset) found the exact distribution of a different statistic, which he denoted by z, namely, $z = (\overline{X} - \mu)/s$, where he defined s^2 as $\Sigma(X - \overline{X})^2/n$. Comparing the t- and z-statistics here, we see that $t = z\sqrt{(n-1)}$.

Several years later, R.A. Fisher, wanting a general test statistic that would unify tests on a single mean, on the difference of two means, on a regression coefficient and on the difference of two regression coefficients, worked out the exact distribution of the t-statistic (exactly as defined above) – Fisher, himself, used the letter 't' to denote the statistic – and published it in 1923. This exact distribution is a function involving a single parameter, which Fisher named 'the degrees of freedom'. In symbols, his result was equivalent to the distribution of $z\sqrt{\nu}$, where ν is the number of degrees of freedom. Whereas Gosset's z-statistic suits only the test on a single mean, Fisher's t-statistic generalises to suit each of the above tests, provided the appropriate numerical value is used for the degrees of freedom.

So, if Fisher, with his *t*-distribution, gave a *complete* solution for a class of hypothesis tests when the population variance is unknown, why is it today called Student's *t*-distribution, rather than Fisher's *t*-distribution?

References

Print

Box, J.F. (1978). *R.A. Fisher: The Life of a Scientist*. Wiley.

David, F.N. (1962). *Games, Gods and Gambling – A History of Probability and Statistical Ideas*. Griffin.

Forrest, D.W. (1974). *Francis Galton: The Life and Work of a Victorian Genius*. Elek.

Hacking, I. (1990). *The Taming of Chance*. Cambridge University Press.

Heyde, C.C. and Seneta, E. (eds, 2001). *Statisticians of the Centuries*. Springer.

Kendall, M.G. (2003). Hiawatha designs an experiment. *Teaching Statistics* **25**, 34–35.

Lehmann, E.L. (1993). The Fisher, Neyman-Pearson theories of testing hypotheses: one theory or two? *Journal of the American Statistical Association* **88**, 1242–1249.

Magnello, M.E. (2009). Karl Pearson and the establishment of mathematical statistics. *International Statistical Review* **77**, 3–29.

McMullen, L. (1939). 'Student' as a man. *Biometrika* **30**, 205–210. Reprinted in Pearson, E.S. and Kendall, M.G. (eds, 1970). *Studies in the History of Statistics and Probability*. Griffin.

Plackett, R.L. (ed, 1990). *'Student' – A Statistical Biography of William Sealy Gosset*. Clarendon Press, Oxford.

Porter, T.M. (2004). *Karl Pearson: The Scientific Life in a Statistical Age*. Princeton University Press.

Reid, C. (1982). *Neyman – from Life*. Springer.

Salsburg, D. (2001). *The Lady Tasting Tea: How Statistics Revolutionized Science in the Twentieth Century*. Freeman.

Stigler, S.M. (1986). *The History of Statistics: The Measurement of Uncertainty Before 1900*. Harvard University Press.

Stigler, S.M. (1999). *Statistics on the Table: The History of Statistical Concepts and Methods*. Harvard University Press.

Student [W.S. Gosset] (1908). The probable error of a mean. *Biometrika* **6**, 1–25.

Weisberg, H.I. (2014). *Willful Ignorance: The Mismeasure of Uncertainty*. Wiley.

Zabell, S.L. (2001). Ronald Aylmer Fisher, pp. 389–397 in Heyde and Seneta (eds, 2001).

Online

[22.1] Galton, F. (1908). *Memories of My Life. Methuen.* At http://galton.org

[22.2] O'Connor, J.J. and Robertson, E.F. (2003). Egon Sharpe Pearson. *MacTutor History of Mathematics Archive.* At http://www-history. mcs.st-andrews.ac.uk/Biographies/Pearson_Egon.html

[22.3] Lindley, D.V. (1980). L.J. Savage – his work in probability and statistics. *Annals of Statistics* **8**, 1–24. At https://projecteuclid.org/euclid. aos/1176344889

[22.4] O'Connor, J.J. and Robertson, E.F. (2010). Leonard Jimmie Savage. *MacTutor History of Mathematics Archive.* At http://www-history. mcs.st-and.ac.uk/Biographies/Savage.html

[22.5] Brillinger, D.R. (2002). John Wilder Tukey (1915–2000), *Notices of the American Mathematical Society* **49**, 193–202. At http://www.ams.org/ notices/200202/fea-tukey.pdf

[22.6] Anscombe, F.R. (2003). Quiet contributor: the civic career and times of John W. Tukey, *Statistical Science* **18**, 287–310. At https:// projecteuclid.org/euclid.ss/1076102417

[22.7] http://www-history.mcs.st-andrews.ac.uk/HistTopics/Statistics.html

[22.8] http://www.york.ac.uk/depts/maths/histstat

[22.9] http://www.economics.soton.ac.uk/staff/aldrich/Figures.htm

[22.10] https://projecteuclid.org/all/euclid.ss

23

Statistical eponymy

We are all familiar with things being named after people. It is very common in geography, with Constantinople (a city), Tasmania (a region), Everest (a mountain) and Victoria (a lake) as examples. It's also very frequent in science, technology, medicine and mathematics, where plants (fuchsia), chemical elements (einsteinium), temperature scales (Celsius), physical laws (Newton's), industrial processes (pasteurisation), diseases (Alzheimer's), mathematical theorems (Pythagoras's), and codes (braille) are often named after people. Such naming of things after people is termed 'eponymy', and the person whose name is used is the 'eponym'. A pleasant excursion through many contexts of eponymy can be found in a 1983 essay by the US information scientist, Eugene Garfield, titled 'What's in a name: the eponymic route to immortality' (online at [23.1]).

A Panorama of Statistics: Perspectives, Puzzles and Paradoxes in Statistics, First Edition.
Eric Sowey and Peter Petocz.
© 2017 John Wiley & Sons, Ltd. Published 2017 by John Wiley & Sons, Ltd.
Companion website: www.wiley.com/go/sowey/apanoramaofstatistics

In this chapter, we shall look at eponymy in statistics. Statisticians' names are attached to concepts, constructs and procedures in every facet of statistical theory. In descriptive statistics we find, for example, the Winsorised mean, Spearman's correlation coefficient and Kendall's concordance coefficient. Many probability distributions carry the name of a statistician or mathematician. Most prominent is the Gaussian distribution (also called the normal distribution), and there are also, among others, the Bernoulli, Poisson, Cauchy, Weibull and Wishart distributions. Many fundamental theorems that underpin the theory of statistical inference are eponymous – for example, the Lindeberg-Lévy Central Limit Theorem, the Neyman-Pearson theorem and the Rao-Blackwell theorem.

Eponymous estimators include the James-Stein estimator (see QUESTION 15.5) and the Horvitz-Thompson estimator. More numerous are eponymous hypothesis tests – for example, Fisher's (exact) test, the Chow test, the Wald test and the Kolmogorov-Smirnov test.

Eponymy has no formal rules and little consistency from case to case, as we note in the answer to QUESTION 22.5 (b). Thus, oddities are to be expected.

There is no fixed time-pattern in the emergence of eponymies. Some appeared shortly after the associated innovation, while others arose only decades later. Examples are found in this chapter's questions.

Some eponymies have faded with time. Pearson's measure of skewness is now more commonly known as the moment measure of skewness (in contrast to the quartile measure of skewness). Snedecor's F distribution (an enhanced presentation by George Snedecor, in 1934, of a construct by R.A. Fisher) is nowadays simply the F distribution. And the Aitken estimator of 1934 is, today, more usually termed the generalised least squares estimator.

In general, the eponym is someone who developed, or materially refined, the construct or procedure in question. However, he or she is not always the earliest among those with competing claims to be the originator. Indeed, Stephen Stigler goes so far as to declare that, regardless of the field, 'no scientific discovery is named after its original discoverer'. Stigler, rather tongue-in-cheek, dubs this dogmatic proposition, 'Stigler's Law of Eponymy'. (The essay with this title, which first appeared in 1980, is reprinted as chapter 14 in Stigler, 1999.)

Of course, if Stigler's Law is true then Stigler is not its originator! Who, then, has that distinction? A claim has been made for Carl Boyer, in his *History of Mathematics* (Boyer, 1968). See page 469 and examples, spread through chapters 18–24, of eponymy awarded to non-originators in mathematics. That claim is documented in the article, 'Who discovered Boyer's Law?' by Kennedy (1972) – see the entry on 'eponymy' (online at [23.2]), contributed by John Aldrich to Jeff Miller's website on the early history of terms in mathematics and statistics. Alas, as Kennedy's title hints, there is now a new regress – for if there is a 'Boyer's Law' then, presumably, Boyer cannot be its originator!

But Stigler is being whimsical. Indeed, he readily concedes that the reader need only 'grant the frequent truth of the Law, and agree to the unreliability of eponyms as guideposts to original discovery'.

Stigler cites several statistical contexts in which his Law, even in its concessional form, is demonstrably true. These include Poisson's statement of what is today called the Cauchy distribution, more than a quarter of a century before Cauchy (see Stigler, 1999, chapter 18), and de Moivre's statement of the formula for the Gaussian distribution in a work of 1733, well before Gauss first referred to it in 1809 (see CHAPTER 14).

One puzzle remains: why are some statistical concepts and techniques accorded eponymic descriptors, while others are not?

In a very few areas of the subject, we observe that eponymy for the originator of a useful method seems to be virtually automatic. These areas include price index numbers, parametric tests for heteroscedastic disturbances in regression models, and non-parametric testing. In other areas where a major innovator is readily identifiable (e.g. Chester Bliss for the probit, and Bradley Efron for the bootstrap), no eponymy has emerged. This only makes the process more intriguing.

Questions

Question 23.1 (B)

The following statistical tests are known by the names of their originators. In each case explain what purpose the test serves and briefly identify the statistician(s) involved:

i. the Behrens-Fisher test;
ii. the Durbin-Watson test;
iii. the Wilcoxon signed-ranks test.

Question 23.2 (B)

Which of the following eponymous statistical constructs was actually originated by the person whose name it is given? If the eponym was not the originator, who was?

i. Chebyshev's inequality;
ii. the Lorenz curve;
iii. Bayes' theorem.

Question 23.3 (B)

The German statistician Ladislaus von Bortkiewicz (1868–1931) is today best remembered for his presentation of a data set that is frequently used to demonstrate the chi-squared goodness-of-fit test to a distribution that was *not* named after him. To what does the data set relate, what is the distribution, and when did it get its name?

Question 23.4 (B)

In 1973, Herman Chernoff published an ingenious idea for representing multivariate data as a simple picture. What sort of picture is this?

Question 23.5 (B)

As a very young man in 1866, and before he became famous as the composer for all the Gilbert and Sullivan operettas, Sir Arthur Sullivan wrote the music for a short operetta with only three singers in the cast. Almost 100 years later, this operetta inspired two eminent statisticians to collaborate in publishing a now well-known statistical paper. What is the title of the paper, who are the authors, and what is the connection with the operetta?

References

Print

Boyer, C. (1968). *A History of Mathematics*. Wiley.

Kennedy, H.C. (1972). Who discovered Boyer's Law? *The American Mathematical Monthly* **79**, 66–67.

Stigler, S.M. (1999). *Statistics on the Table: The History of Statistical Concepts and Methods*. Harvard University Press.

Online

[23.1] www.garfield.library.upenn.edu/essays/v6p384y1983.pdf

[23.2] http://jeff560.tripod.com/e.html

24

Statistical 'laws'

When people speak of 'the law of gravity', they are generally referring to what is more exactly called 'Newton's Law of Universal Gravitation'. This law states that the gravitational force (that is, the mutual attraction) between any two physical bodies is directly proportional to the product of their individual masses and inversely proportional to the square of the distance between them.

Why would such a scientific relationship be called a 'law'? An analogy, while imperfect, may be helpful. Think about the word 'law' as it is used in parliament.

A law is a rule of behaviour that parliament has agreed is binding on people everywhere in society. Parliamentarians agree on what behaviour should become law only after having clear evidence of the expected social benefits of the law. Similarly, a physical law is a rule of behaviour that scientists have agreed to regard as binding on physical matter everywhere in nature. Scientists agree on what behaviour of matter should be called a law only after having clear evidence of its *major scientific importance*.

To this italicised characteristic of a scientific law, we can add five more. When a scientific law represents a relationship between variables, that relationship *can be expressed in simple terms*: it relates the 'response' variable to just a few 'stimulus' variables. The relationship is usually *causal*: it implies not only a correlational connection between the stimulus variables and the response variable, but also a direct determining mechanism (for more on correlation and causation, see the answers to QUESTIONS 8.2 and 9.1). The relationship is *stable* – that is, the determining mechanism is unchanging over time and/or place. And because those who are able to identify such simple and stable causal relationships in an otherwise complex

A Panorama of Statistics: Perspectives, Puzzles and Paradoxes in Statistics, First Edition.
Eric Sowey and Peter Petocz.
© 2017 John Wiley & Sons, Ltd. Published 2017 by John Wiley & Sons, Ltd.
Companion website: www.wiley.com/go/sowey/apanoramaofstatistics

and turbulent world are, for that reason, quite remarkable scientists, scientific laws are mostly named in their honour – that is, they are *eponymous*, as the name 'Newton's Law' illustrates (see CHAPTER 23 for more on eponymy).

Finally, because there can be no formal proof that a scientific law is universally true, even long-established scientific laws are always vulnerable to being shown to be only approximations. In other words, they *may need modification* as observation becomes more acute, measurement becomes more accurate, and confirmatory experiments are conducted in more unusual or extreme situations. Newton's Law is, again, a good example. Newton's account of gravitational attraction implies that this force operates instantaneously, regardless of distance. This suffices as an excellent basis for Earth-bound physics. However, this notion was contradicted by Einstein's Theory of Relativity, a theory now empirically well confirmed over interplanetary distances.

Many other eponymous physical laws were established prior to 1900, including Boyle's Law, Ohm's Law, Hooke's Law, and Kepler's Laws. The 20th century was an era of huge growth in the social and behavioural sciences. It was natural, then, for scholars to ponder whether there are laws in these sciences, too. One way they could seek an answer was to search empirically for 'law like' relationships (that is, simple and stable relations among variables), using statistical methods.

Of course, a strong statistical correlation, together with a stable regression model, does not necessarily signal that a direct causal mechanism has been identified. However, it certainly is a constructive first step in that direction. Thereafter, one can theorise about a plausible general causal mechanism to explain the stability of the statistical findings. Just as important, one can map out the limits beyond which the causal mechanism is not expected to apply. In this way, a new law may be tentatively proposed, to be subjected to further tests for confirmation. Examples of this approach are given in Ehrenberg (1968).

---oOo---

Not all scientific laws represent relationships between variables. There are also statistically discovered laws that relate to the frequency distribution of just a single variable. It turns out that, for certain measured variables, the relative frequency of their repeated measurement in the real world is very well approximated by some standard probability model (see CHAPTER 13 for more on probability models). Indeed, that is precisely why such probability models became 'standard'!

If a model's fit remains close when applied to repeated measurement data on a particular variable collected in widely different settings, statisticians may 'promote' the model to the status of a *probability law* for that variable.

Take, for instance, the 'random error of measurement' of some fixed quantity. This is the variable for which the first probability law in the history of statistics was designated. It was Gauss who, in 1809, first proposed the probability model; later, it became known as the 'normal law of error'. It was soon well confirmed that repeated measurement of some fixed quantity produces a roughly symmetric distribution of random measurement errors, x, which is well approximated by a normal probability distribution of the form:

$$f\left(x \,|\, \mu = 0, \sigma\right) = \frac{1}{\sigma\sqrt{2\pi}} \exp\left(-\frac{1}{2}\left(\frac{x}{\sigma}\right)^2\right)$$

where the parameter σ is the population standard deviation. We note that, in this context, it is reasonable to set the parameter μ (the population mean) to zero, since errors are equally likely to be positive or negative.

When Gauss proposed the normal as a probability model, its elaborate form must have astonished many. It can equally astonish beginning students of statistics today. How could such an unobvious and abstruse mathematical function (they must wonder) ever have been hit upon? That seems to us a perfectly understandable reaction, if students are introduced to the mathematical function without any background.

It will be helpful background if students come to see that the function we know as the normal distribution was not plucked out of the air. It was already known to Gauss from work by de Moivre, several decades earlier, on the limit of the binomial distribution as the sample size increases without limit. Gauss favoured it in 1809 precisely because – as we mention in CHAPTER 14 – it is the only symmetric continuous probability density function for which the mean of a random sample has the desirable property of being the maximum likelihood estimator of the population mean. The normal also underpins appealing statistical properties of many statistical tools, including point and interval estimators, and significance tests.

However, the normal probability model has its limitations! Though it has served statisticians superbly for over 200 years, the normal is not necessarily the best probability model for every symmetrically distributed unimodal variable.

Nor is it necessarily the best probability model for *the mean of a large sample* drawn randomly from any non-normal population – despite what the Central Limit Theorem (CLT) promises. This powerful theorem (explained in CHAPTER 12) greatly widened the scope of the normal as a probability model after it was first established in 1809. Yet, the CLT has its limitations, too!

---oOo---

Let's look now at a variable that will lead us to a probability law which is *not* the normal distribution. This is a law that has become increasingly significant over the past 50 years.

Since the 1960s, there have been many statistical studies of Stock Exchange data. Among the statistics studied was the daily average of relative price changes, over all the shares in the category labelled 'speculative' (i.e. shares liable to frequent strong stochastic shocks to their prices). It was soon noticed that the frequency distribution of these *average short-term relative price changes* had many extreme values, both positive and negative. These empirical distributions were unimodal, and roughly symmetrical about a mean of zero, but the frequency in their tails was greater than a normal distribution would imply. In other words, the tails of the empirical distributions were 'fatter' (or 'heavier') than those of the normal (QUESTION 14.1 (c) shows just how 'thin' are the tails of the normal).

At first, attempts to model the mean of relative price changes proceeded by treating the extreme values as (alien) outliers, deleting these outliers from the data set and fitting the normal as a probability model to the central data values. Case-by-case explanations were then contrived, to account for the size and frequency of the outliers. The results of this piecemeal approach were not very satisfactory.

In 1963, Benoit Mandelbrot (1924–2010), a French-American mathematician, proposed a new approach to the modelling challenge. He drew on a family of probability distributions identified by the French mathematician Paul Lévy (1886–1971), known as 'stable distributions'. We have more to say about Mandelbrot's work shortly. First it is useful to have a brief look at what exactly a stable distribution is, and what Lévy found out about this family.

Lévy's explorations in this area were a by-product of his research, over the years 1920–1935, on proofs of the CLT under progressively relaxed conditions. Recall, from CHAPTER 12, that the CLT says 'if you draw a sample randomly from a population that is *not* normally distributed, the sample mean will nevertheless be approximately normally distributed, and the approximation will improve as the sample size increases'. We note that there is one condition that cannot be relaxed. The CLT is valid only for a population with a *finite variance*.

Means of random samples from a *normal* population have, of course, a normal distribution for *every* sample size. One might see this property of normal means as a sort of 'trivial' case of the CLT. But Lévy saw it differently! What he focused on was that here we have a case where *the sample means have the same form of distribution as the individual sample values for every sample size*. That made him wonder whether there might be a definable family of probability distributions that all have this property of 'stability'.

Other mathematicians (Poisson and Cauchy in the 19th century and Pólya, Lévy's contemporary) had discovered some individual members of the family, but it was Lévy who (in two short papers in 1923) elegantly characterised the entire family of stable distributions, both symmetric and non-symmetric. Today, his results can be found in many advanced textbooks of probability and statistics.

Here is a sketch, without mathematical derivations, of some attributes of the family of stable probability distributions that Lévy discovered.

All the stable distributions are unimodal. They are unified by a 'characteristic' parameter (call it α) which lies in the range $0 < \alpha \le 2$. Stability is defined in a quite specific sense: if sample values are all drawn independently from the same stable distribution – say, the distribution with characteristic parameter value α^* – then the sample mean will have the distribution with characteristic parameter value α^* at every sample size.

Only two *symmetric* stable distributions have an explicit form of probability density function (pdf): the normal (corresponding to $\alpha = 2$) and the Cauchy (corresponding to $\alpha = 1$). For all others, a probability is defined formally in terms of the convergent sum to infinity of a rather forbidding algebraic expression. In practice, these probabilities are calculated by evaluating that sum up to any desired level of accuracy.

None of the stable distributions (except the normal) has a finite variance. It follows that none of the stable distributions with α in the range $0 < \alpha < 2$ conforms to the CLT.

Lastly, here is the property that made the stable distributions so particularly interesting to Mandelbrot: all the distributions with α in the range $0 < \alpha < 2$ have fatter tails than the normal.

If you would like to read about the mathematics of stable distributions, we suggest you start with the accessible account of basic ideas in Borak *et al.* (2010), online at [24.1]. A rather more advanced, yet invitingly written, overview is given in chapter 9 of Breiman (1992).

---oOo---

Let us pause for a moment to look at the Cauchy distribution – a symmetric stable distribution that is perhaps less familiar to you than the normal. The Cauchy distribution has no finite mean or variance. That explains why its two parameters – a measure of centrality and a measure of spread – are positional measures. Its pdf is:

$$f\left(x \mid \theta, \lambda\right) = \frac{1}{\pi\lambda}\left[1 + \left(\frac{x-\theta}{\lambda}\right)^2\right]^{-1}$$

where $\pi = 3.14159...$, θ is the median and λ is the semi-interquartile range. The standard Cauchy distribution is given by setting $\theta = 0$ and $\lambda = 1$. Its pdf is $f(x) = [\pi(1 + x^2)]^{-1}$.

The standard Cauchy distribution is graphed together with the normal distribution having the same median and semi-interquartile range ($\mu = 0$, $\sigma = 1.4827$) in FIGURE 24.1.

You can see that beyond about ± 3, the Cauchy has fatter tails than the normal. This is the property that makes the Cauchy more useful than the normal for modelling financial data having multiple extreme values.

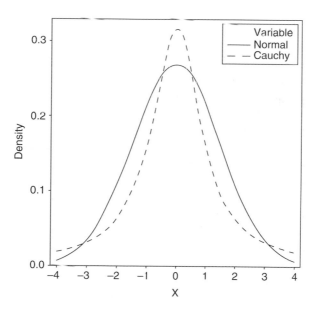

Figure 24.1 Graphs of the density functions of the normal and Cauchy distributions.

---oOo---

It was not until 40 years after Lévy announced the entire family of stable distributions – the first class of fat-tailed distributions to be identified – that the importance of this theoretical work was recognised in applied research. As already mentioned, it was the study by Mandelbrot (1963) that first demonstrated the need to accommodate the fat tails of empirical distributions of average short-term relative changes in share prices. A better fitting model than the normal distribution was needed. Since the fat-tailed stable distributions were to hand, Mandelbrot tried them out; it was their pioneering empirical role.

In his 1963 article, Mandelbrot, a mathematician, writes mostly in algebraic terms about the modelling issues he confronted. You may find it easier to read Fama (1963). Eugene Fama, an economist and Mandelbrot's younger colleague, commends the practical significance of Mandelbrot's ideas to the journal's readership of economists. Fama continued in this research direction (see, for example, Fama, 1965), even after Mandelbrot turned to different topics (most famously, the geometry of fractals). In 2013, Fama shared the Nobel Prize in Economics, in part for the lasting impact that these successful early studies have had on the evolution of financial mathematics and statistics.

Since this early work on modelling aspects of speculative share prices, many other risk-related financial variables, including some that are non-symmetrically distributed, have been found to have fat-tailed distributions. These, too, have been effectively modelled by members of the family of stable distributions.

To this literature can be added a remarkably extensive array of stable models of fat-tailed variables in physics, geology, climatology, engineering, medicine and biology. There is also an array of multiple regression models where there are grounds for assigning the random disturbance a non-normal stable distribution, rather than the more usual normal distribution.

With so many fat-tailed variables, in such diverse contexts, well modelled by stable distributions, there is ample evidence for 'promoting' them to the status of *stable laws*.

There are, however, fat-tailed variables – especially non-symmetric ones – for which the stable laws do not provide a well-fitting model. For these cases, there are now several other theoretical distributions which may be deployed instead. They include the lognormal distribution, the generalised hyperbolic distribution, the geometric distribution, or one of the power laws investigated in QUESTION 24.3.

For the important statistical law called the 'law of large numbers' and two *misconceived* laws – the 'law of averages' and the 'law of small numbers' – see the answer to QUESTION 3.1.

Questions

Question 24.1 (B)

The eminent French mathematician, probabilist, engineer and philosopher of science Henri Poincaré (1854–1912) taught at the Ecole Polytechnique in Paris at the peak of his career. At this time, he published a highly-regarded textbook of probability (Poincaré, 1896). In chapter 10 of this book, Poincaré demonstrates algebraically how Gauss, in 1809, first obtained the probability density function of the normal law of error. To lighten the detailed mathematics, Poincaré interpolates an anecdote about himself and a colleague, the physicist Gabriel Lippmann. Perhaps they had been discussing the scientific community's lack of interest in the true nature of disciplinary foundations, for Poincaré writes, about the normal law of error, 'Everyone believes in it, Mr. Lippman once told me, since empiricists suppose it's a mathematical theorem and mathematicians that it's an experimentally determined fact.'

Which of these, would you say, is the true nature of the normal law of error?

Question 24.2 (B)

In a large collection of (say, five-digit) random numbers, you would expect that the digits 1, 2, 3, 4 … 9, 0 would turn up with roughly equal frequency as the leading (that is, first) digit of those numbers. Surprisingly, however, in many real-world collections of numbers (for example, the serial numbers of business invoices or the money amounts on electricity bills), some leading digits are actually more likely than others (zero is excluded as a leading digit). This phenomenon was stumbled upon in the 19th century from an incidental observation that earlier pages of books of logarithms were more 'worn' than later pages.

What probability distribution is applicable, in these circumstances, for the frequency of occurrence of leading digits? How does knowledge of this distribution assist auditors looking for accounting fraud?

Question 24.3 (B)

Newton's Law of Gravitation is an instance of a power law. Many physical, economic and social variables have a frequency distribution that follows a power law. What, in general, is a power law?

It has been found nowadays that aspects of website traffic on the internet follow a power law. Give an example in this context.

Question 24.4 (C)

A particular power law that relates to discrete data is Zipf's Law. This was initially proposed as a probability law for the relative frequency with which individual words appear in an extended prose text, regardless of language. Who was Zipf, and what is the functional relationship that bears his name?

Subsequently, Zipf's Law has been found to apply in many other situations as well, notably the rank-size distribution of the cities in any particular country. For your country, write down at least the first 15 major cities, in order by population size (measured in thousands), ranking the largest city 1, the next largest 2 and so on. Next, create a scatter diagram of the data, with the X-axis showing the logarithm of population and the Y-axis the logarithm of the city rank. What do you see? If the software is available, fit a least squares regression line to the scatter. What do you find? Relate your finding to Zipf's Law.

Question 24.5 (B)

In the 1950s, a French husband-and-wife team of psychologists/statisticians published their discovery of a remarkable phenomenon, that famous sports people are statistically significantly more likely to have been born when the planet Mars was in particular positions in the sky – the so-called 'Mars effect'. Who were they? Has their discovery a record of confirmation since then that would justify giving it the status of a law?

References

Print

Breiman, L. (1992). *Probability*. Society of Industrial and Applied Mathematics (this book is a reprint of the original Addison-Wesley publication of 1968).

Ehrenberg, A.S. (1968). The elements of lawlike relationships. *Journal of the Royal Statistical Society, Series A* **131**, 280–302.

Fama, E.F. (1963). Mandelbrot and the stable Paretian hypothesis. *The Journal of Business* **36**, 420–429.

Fama, E.F. (1965). The behavior of stock market prices. *The Journal of Business* **38**, 34–105.

Mandelbrot, B. (1963). The variation of certain speculative prices. *The Journal of Business* **36**, 394–419.

Poincaré, H. (1896). *Calcul des Probabilités*, Gauthier-Villars (2nd edition in 1912).

Online

[24.1] Borak, S., Misiorek, A. and Weron, R. (2010). *Models for heavy-tailed asset returns*. SFB649 Discussion Paper 2010-049, Humboldt University, Berlin. Downloadable at http://hdl.handle.net/10419/56648. (*Note*: this paper also appears as chapter 1 in Cizek, P., Härdle, W. and Weron, R. (eds, 2011). *Statistical Tools for Finance and Insurance*, 2nd edition. Springer).

25

Statistical artefacts

The principal meaning of the word 'artefact', as given in the Macquarie Dictionary, is 'any object made by man with a view to subsequent use'. If we stretch this definition somewhat, the earliest *statistical* artefact was surely the hand. Humans have been counting – a proto-statistical task – since Stone Age times. From the first, it is likely that counting was performed using various parts of the body, especially the fingers and toes, and this is still evident today in the traditional counting systems of some groups of people. A lasting record of a count could be produced by making notches on bones or sticks.

Also very old, but much more sophisticated as a counting record, is the yoked set of knotted strings, called a *quipu*, of the Inca civilisation in South America. In this chapter, we investigate the quipu as a statistical artefact, and follow with an overview of some other kinds of statistical artefacts.

Quipu is an unusual word in statistics, and one of very few that begin with the letter 'q' (see QUESTION 25.3 for another one!). In fact, quipu is the Spanish spelling of the word and also the most common spelling in English. However, an alternative spelling that seems to be gaining prominence in anthropology is khipu, which is taken from the Quechua language. Quechua was the official language of the Inca Empire of the 15th and 16th centuries, and is still a living language today. Since the Spanish conquest of Peru in the mid-16th century, Quechua has been written in the Latin alphabet. The khipu spelling is a better phonetic guide to the correct pronunciation, which has an initial aspirated *kh*, rather than the *kw* that the spelling *quipu* suggests to an English speaker.

So, what type of statistical artefact is a khipu? Its knotted strings are usually of llama or alpaca wool. Numerical and, possibly, qualitative information is encoded in the number and type of knots, and also in the colour and weave

A Panorama of Statistics: Perspectives, Puzzles and Paradoxes in Statistics, First Edition.
Eric Sowey and Peter Petocz.
© 2017 John Wiley & Sons, Ltd. Published 2017 by John Wiley & Sons, Ltd.
Companion website: www.wiley.com/go/sowey/apanoramaofstatistics

of the strings. Khipus record economic, financial and demographic information. It seems that they contain, as well, a location indicator for the information source. They were used to convey information back and forth between
a local area and a central administration. They were created and interpreted
by khipukamayuqs (or quipucamayocs) – that is, khipu experts – or, might
we say, Inca statisticians!

You can see images of many of the 600 or so surviving khipus on the web.
A particularly good site is [25.1], maintained by Carrie Brezine and Gary
Urton of the Khipu Database Project at Harvard University. Ms Brezine is a
mathematician and a weaver and Dr Urton is a professor of Pre-Columbian
Studies. The gallery at their website includes a 'calendar khipu', which is
thought to record statistical information from a two-year period.

On a visit to the *Museo Chileno de Arte Precolombino* (online at [25.2]), in
Santiago, Chile in 2007, one of us (PP) saw a splendid example of a khipu,
displayed with the primary cord in a curve and the several hundred pendant
strings radiating from their points of attachment (see the image online at
[25.3]). He was keen to photograph it, yet photography in the museum was
banned. After he explained in hesitant Spanish that he was 'un profesor de
estadística', the staff, maybe puzzling over the relevance of this declaration
to his request, allowed him to take a photo. Perhaps he should simply have
introduced himself as a khipukamayuq!

We should note that khipus are not solely ancient artefacts, nor are they
found only in South America. Ifrah (1998) devotes chapter 6 of his fascinating book *The Universal History of Numbers* to a discussion of 'numbers on
strings'. He points out that relatives of the khipu are found in a variety of
historical settings, including the Roman Empire and ancient China. Further,
such artefacts are still used today by people in Bolivia and Peru, various
Pacific Islands, Tibet and parts of Africa, to keep records of sizes of livestock
flocks, amounts of money owed, and numbers of prayers recited.

Among many other statistical artefacts, the most powerful today are very
familiar – the computers and computer programs that are used to carry out
most statistical work, namely, data entry and organisation, graphical display
of data, simple and complex statistical analyses, and even manipulation of
complicated algebraic formulae in advancing statistical theory. It is an interesting connection between the old and the new that, in the Quechua version
of *Windows*, the word Kipu (spelled without an 'h') is used for 'file'!

---oOo---

It is a small step to extend the definition of an artefact beyond physical
objects to abstract ideas. Indeed, the Oxford English Dictionary (OED)
includes, as a derivative sense of the word, the definition 'a non-material

human construct'. In statistics, we utilise a whole range of artefacts in this abstract sense, as well – for instance, the normal distribution and the normal curve (though we might think of a printed table of areas under the normal curve as a physical artefact – see CHAPTER 14).

A further meaning of the word artefact is 'a spurious result, effect, or finding in a scientific experiment or investigation, especially one created by the experimental technique or procedure itself' (to quote the OED again). In this distinct sense, too, there are statistical artefacts, and it is essential for us, as statisticians, to recognise them for the 'mirage' that they are, and not be misled into thinking that they represent something authentic.

There are broadly two categories of statistical artefacts of this kind.

The first category groups together the results of inappropriate uses of statistical tools in practice. A particular use may be inappropriate, either because the tool is *theoretically not designed* for that use, or because it is simply *unsuited* to that use.

In our discussion of price indexes in CHAPTER 7 – especially QUESTION 7.1 – you can find a striking example of the arithmetic mean in use in a context for which it is *not designed*. There is another example in QUESTION 18.2, this time featuring the coefficient of determination.

QUESTION 1.3 provides an example of the arithmetic mean applied where it is *unsuited*. There is a further example of an artefact appearing where a tool is unsuited to the context in CHAPTER 18. The artefact is 'regression towards the mean', which commonly arises in regression analyses of repeated measures data.

The second category of statistical artefacts is computational. Long chains of computer calculations are routine in statistical analyses, and it is well known that rounding error can then become quite serious.

An easily appreciated example of the consequences of rounding error is the calculation of a variance. Two alternative algorithms for finding the (population) variance of N values of a variable X, with population mean μ, are:

a) $\quad \Sigma\left(X_i - \mu\right)^2 / N$

and

b) $\quad \left(\Sigma X_i^2 - N\mu^2\right) / N$

In principle, (b) is more susceptible to rounding error than (a), because it involves the small difference of two large numbers. If the large numbers are both computed with insufficient accuracy, their difference may be seriously in error. This is unlikely to happen if (a) is used.

Weisberg (1985), page 282, gives an instructive demonstration of the consequences of choosing algorithm (b) rather than (a). Suppose you are using a calculator which preserves seven significant digit accuracy to find the variance of 12537, 12541 and 12548. The mean of these three values is 12542. Then, using (a), the variance is $(5^2 + 1^2 + 6^2)/3 = 62/3$.

What happens if (b) is used? The values of X_i^2, *rounded to seven significant digits*, are, respectively, 157,176,400, 157,276,700 and 157,452,300.

Then, $\sum X_i^2 = 471,905,400$.

The value of μ^2, *rounded to seven significant digits*, is 157,301,800.

Then $N\mu^2 = 471,905,400$.

Thus, the numerical result of this variance calculation is 0. Of course, this result is a computational artefact!

While it is instructive, this example is hardly realistic by current computing standards. After all, the result of a single arithmetic operation in today's standard 'quadruple-precision binary' representation of numbers preserves about 33-digit accuracy when converted to the equivalent decimal value. However, when you consider that a multivariate statistical analysis may involve tens of thousands of arithmetic operations, and that even tiny initial rounding errors propagate cumulatively through successive rounds of computation, then even quadruple-precision binary arithmetic may not save us from computational artefacts.

Questions

Question 25.1 (A)

Thinking of our hands as statistical artefacts, we have known since childhood how to use them to count from 1 to 10. A mediaeval English monk devised a way of using finger counting to represent much larger numbers. Who was the monk, and how far did his counting system extend?

Question 25.2 (A)

At one of the early censuses in the United States in the 19th century, an important artefact for processing the data was first introduced. What was this artefact and who was its inventor?

Question 25.3 (A)

What sort of an artefact is a quincunx? What would a statistician use it for? Which British statistician brought it into use?

Question 25.4 (A)

Dice are probabilistic artefacts and have been used for millennia in gambling. In the 18th century, however, they were put to a surprising purpose. A famous classical composer wrote a collection of musical fragments and gave instructions for composing a dance from some of these fragments, selected according to the results of several rolls of two dice. Who was the composer and what type of dance resulted?

Question 25.5 (B)

A striking photograph of a histogram, composed of people arranged in groups by their height, was published in 1975 and has been reproduced many times since. Who was the author of the 1975 publication, and what name did he give to such a histogram? Was this the first time such a histogram had been created and photographed?

References

Print

Ifrah, G. (1998). *The Universal History of Numbers*. Harvill Press.
Weisberg, S. (1985). *Applied Linear Regression*, 2nd edition. Wiley.

Online

[25.1] http://khipukamayuq.fas.harvard.edu
[25.2] http://chileprecolombino.cl/
[25.3] http://chileprecolombino.cl/coleccion/quipu/

Part VI

Answers

26

Answers to the chapter questions

Answers – Chapter 1

Question 1.1

a) John Edmund Kerrich (1903–1985) was born in England, and emigrated to South Africa in 1904. After tertiary studies in South Africa and the UK, he joined Witwatersrand University as a Lecturer in Mathematics in 1929, being promoted to Senior Lecturer in 1935. In 1934, he married a woman of Danish parentage, and was on a visit to her family in Copenhagen in 1940 when the Nazi invasion of Denmark occurred. He was interned by the Nazis in a prison camp in Jutland from 1940 to 1945. It was as a pastime during this period of internment that he performed the 10,000 coin tosses and other experiments for which he has become renowned in the statistical literature.

  After the War, he published the results of these experiments in a short book (Kerrich, 1946). Reviews can be found in the *Journal of the Royal Statistical Society, Series A* of 1947 (page 74) and the *Journal of the American Statistical Association* of 1949 (page 147). The book was apparently later reprinted by the University of Witwatersrand Press. Kerrich became the Foundation Professor of Statistics at Witwatersrand University in 1957, retiring in 1971. A short biography, including a photo, was published in the *South African Statistical Journal*, 7(2), 1973, 82–83.

b) We prepared FIGURE 1.1 from Kerrich's original data, as given on page 274 of Freedman, Pisani and Purves (2007). Over all the 10,000 coin tosses, these data show the relative frequency of heads to be 0.5067.

A Panorama of Statistics: Perspectives, Puzzles and Paradoxes in Statistics, First Edition.
Eric Sowey and Peter Petocz.
© 2017 John Wiley & Sons, Ltd. Published 2017 by John Wiley & Sons, Ltd.
Companion website: www.wiley.com/go/sowey/apanoramaofstatistics

If we choose to define probability as 'relative frequency of occurrence', then we may say that *an approximation to* the probability of a head *for the particular coin that Kerrich used* in his experiment is 0.5067. Why is it just an approximation? It is clear that the empirical relative frequency of a head changes with each additional set of tosses. How can any one of these figures be said to be *the* (true) probability of a head for Kerrich's coin? So how, then, might we define the true probability for that coin?

One mathematically appealing solution is to define that probability as 'the limit of the relative frequency of a head as the number of tosses approaches infinity'. This may be mathematically appealing, but it is hardly edifying in practice. That is because, if we are asked 'what is the numerical probability of getting a head with Kerrich's coin?', the honest answer in the context of this rather abstract definition of probability can only be 'we don't know'!

To get a better understanding of another difficulty with this abstract definition of probability, look again at the graph of Kerrich's empirical data. What it shows is that the deviation from 50% of the cumulative percentage of heads oscillates as the total number of tosses increases, and that it appears to be a damped oscillation. The graph suggests that 'in the long run', the cumulative percentage of heads will converge to some value close to 50%. Statisticians call this apparent convergence 'statistical regularity'. Why does it happen? The best answer to this that we have heard is 'that is how the world is'. We have no proof that there will actually be a convergence – or, in other words, that the limit mentioned in the previous paragraph exists. Therefore, we should be cautious about any definition of probability that depends upon it.

Evidently, getting a clear-cut definition of the probability of a head with Kerrich's coin is mired in complications.

To get away from the problems that arise in defining probability empirically with regard to a particular coin, statisticians have created an abstraction – a 'fair coin'. A fair coin is one whose two faces are perfectly in balance. This idealisation is complemented by another abstraction – a 'fair toss'. A fair toss is one in which neither side is favoured when the coin falls, and the possibility of any outcome other than a head or a tail is excluded (see, for contrast, QUESTION 13.1). Assuming fair tosses of a fair coin, it becomes much easier to define the probability of getting a head. Since the outcome of such an idealised experiment can only be a head or a tail, and they are equally likely outcomes, then the probability of a head must be exactly 0.5. Has Kerrich's experiment provided any support for this 'equally likely' approach to defining the probability of a head? Only to the extent that this approach has been found to be closely consistent with Kerrich's empirical result.

You may recognise, in this brief discussion of three approaches to defining the probability of a head, the ideas of three of the pioneers of probability theory – respectively, Bernoulli, Kolmogorov and Fermat. The fact that these alternative approaches still coexist today (together with a yet a fourth one, namely the subjective approach associated with Savage – see CHAPTER 20), more than 350 years after Blaise Pascal and Pierre de Fermat first felt their way in quantifying chance (see CHAPTER 10), is an indication of how complex a concept is probability. Even though we do not have one unanimously agreed definition of 'the probability of a head', we may perhaps draw comfort from the fact that the three alternative definitions we have considered lead to much the same numerical values, and not worry too much about their variation. (For a parallel conclusion about the concept of randomness, see the answer to QUESTION 11.3.)

Question 1.2

When two dice are rolled, the sample space contains 36 outcomes, taking into account the order in which the spots appear. These 36 outcomes are the entries in the cells of a 6×6 square, in which all couplings of a 1-spot up to a 6-spot on one die are teamed with a 1-spot up to a 6-spot on the other. When young children are asked to say what the sample space is in this problem, they often do not think to take account of the order in which the spots appear. Their answer thus includes only the outcomes in the cells in the lower triangle (below the diagonal) of the 6×6 square just described, plus the outcomes on the diagonal itself. The number of outcomes is then $1 + 2 + 3 + 4 + 5 + 6 = 21$. Omitting to take into account the order in which outcomes may appear is also common among adults. Indeed, it is a centuries-old oversight – see QUESTION 11.1.

It is not incorrect to list only these 21 points as the sample space, but there is a disadvantage. By contrast with the 36 points, the 21 points are not equally likely and, hence, their probabilities cannot be calculated using the 'equally likely' approach to defining probability.

Question 1.3

Most people in London have two legs, but a few have one leg and even fewer have no legs. If the average (of the number of legs per person) is the *median* or the *mode*, both of which have the value 2 in this context, then there are no surprises – most people in London have the average number of legs. However, if the average in question is a *mean*, then we get the paradoxical – but logically correct – result that most people in London have more than the average number of legs. That is because the mean number of legs per person, calculated over all the people in London, will be a number very close to, but smaller than, 2.

How can we explain this paradox? By pointing out that the mean is an inappropriate average in this situation. This does not necessarily signal misuse of statistics, if 'misuse' is understood as implying a deliberate intention to deceive. However, it reminds us to be always vigilant when assessing the conclusions of statistical analyses done by others.

(If you are not familiar with the notion of a paradox, see CHAPTER 10.)

Question 1.4

The monthly average temperature bar charts for New York and New Delhi have similar profiles But, if you look closely, you will see that between July and September, New Delhi is only a little hotter on average than New York. In the remaining months, New Delhi is clearly the hotter city.

To give an informative account of the relative temperatures in these two cities, you should not limit yourself to a comparison of whole-of-year *average* temperatures. A better picture will emerge if you also contrast the relative *spreads* of monthly temperatures over the year in these cities. To generalise, two data distributions are more informatively compared numerically by comparing their means and their standard deviations, rather than by comparing their means alone (see CHAPTER 6). This principle is even more vividly illustrated in a comparison of monthly temperatures in New Delhi and Singapore. While these cities have essentially the same whole-of-year mean temperature, the spreads of monthly temperatures across the year are strikingly different!

Note: the question wording suggests that New Delhi (average temperature 25.2°C) is roughly twice as hot as New York (average temperature 11.7°C). If you think this is a reasonable interpretation of the data, see the answer to QUESTION 8.5 to dispel that thought.

Question 1.5

Outbreaks of cholera occurred in London irregularly from the 1830s to the 1850s, causing massive public health crises. Without an understanding of the epidemiology of cholera, doctors could treat their patients only symptomatically. In that era, long before the advent of antibiotics, severe infection with the cholera bacterium generally proved fatal.

When cholera struck in central London's Soho district in 1854, John Snow (1813–1858), a London medical practitioner, investigated the locations of homes where cholera deaths had occurred. His previous experience with cholera had led him to the hypothesis that it was carried by impurities in water. Snow found a dramatic clustering of deaths around a particular street water pump – the Broad (today called Broadwick) Street pump, located near the corner of what is, today, Lexington Street. Most homes in that area had no alternative domestic water supply. Therefore, by disconnecting this pump and noting that the number of new cholera cases locally dropped off markedly, Snow had strong support for his hypothesis. This event laid a foundation for research on polluted drinking water to identify what exactly caused cholera infections.

It has long been considered that Snow was led inductively from his map of cholera deaths to postulate the Broad Street pump as a major source of the infection. A fascinating account along these lines is given in chapter 3 ('Snow on Cholera') of Goldstein and Goldstein (1978). However, Brody *et al.* (2000) offer evidence for their belief that Snow's hypothesis came first, and was tested by investigating the location of deaths, and that the map was drawn only for his final report in December 1854. The picture we present in FIGURE 1.2 comes from Brody *et al.* It is a small part of Snow's map, showing the position of the Broad Street pump in relation to nearby cholera deaths, coded as parallel dark dashes.

The internet is rich in documents on John Snow. A particularly comprehensive site is maintained by the School of Public Health at the University of California, Los Angeles (see [1.2]). The Brody article just mentioned is also included there, at [1.3].

Snow's map was one of the first half dozen efforts of this kind to track down the source of epidemic diseases in Britain, as Gilbert (1958) describes. Because Snow went on to research in greater depth the relation between the spread of cholera and the contaminated water supply, he is sometimes spoken of as the founder of the field of epidemiology in Britain, if not also worldwide.

References

Print

Brody, H., Rip, M.R., Vinten-Johansen, P., Paneth, N. and Rachman, S. (2000). Map-making and myth-making in Broad Street: the London cholera epidemic 1854. *The Lancet* **356**(9223), 64–68.

Freedman, D., Pisani, R. and Purves, R. (2007). *Statistics*, 4th edition. Norton.

Gilbert, E.W. (1958). Pioneer maps of health and disease in England. *Geographical Journal* **124**, 172–183.

Goldstein, M. and Goldstein, I. (1978). *How We Know: An Exploration of the Scientific Process*. Plenum Press.

Kerrich J.E. (1946). *An Experimental Introduction to the Theory of Probability*. Einar Munksgaard, Copenhagen.

Online

[1.2] (1946b). http://www.ph.ucla.edu/epi/snow.html
[1.3] (1946c). http://www.ph.ucla.edu/epi/snow/mapmyth/mapmyth.html

Answers – Chapter 2

Question 2.1

a) Applying Pythagoras's Theorem, the mathematician states that the length of the diagonal is $20\sqrt{2}$ cm. This is the accurate (i.e. true) length of the diagonal. However, it is an abstract representation: $\sqrt{2}$ is not a scale mark on any measuring tape. Given that $\sqrt{2} = 1.41421...$, you can represent the length in concrete numerical terms as 28.0 cm or 28.20 cm or 28.280 cm, depending on how many decimal places you use to progressively approximate $\sqrt{2}$. These are all inaccurate statements of the length, but they are successively more accurate approximations.

b) The statistician knows that the arithmetic mean of a random sample of length measurements is, on several criteria, a good estimator of the true length. (Think of the true length here as the mean of the population of 'all possible' careful measurements of the diagonal. 'All possible' implies that this population is infinitely large; it is a notional population, rather than a countable one.)

Suppose a sample of 25 measurements of the square's diagonal, recorded to the nearest tenth of a millimetre by 25 randomly chosen people, shows a mean length of 28.26 cm. Then 28.26 cm is the statistician's (point) estimate of the length of the diagonal. This procedure may, or may not, produce an accurate statement of the length (in this particular case, you can see, from part (a), that it is, in fact, inaccurate). However, no other way of combining the information in the 25 measurements is sure, in general, to give a single number that is a more accurate approximation to the true length.

While a point estimate provides no insight as to how accurate it is, an interval estimate is more constructive. Thus, if the 25 measurements are used to derive a 95% confidence interval for the length of the square's diagonal, then one may act in practice as if this interval contains the true length, since such an interval, calculated repeatedly over a long run of samples of 25 measurements, has a 95% success rate in capturing the true length. Note that a confidence interval can be obtained, whether the sample data come from a finite or an infinite population.

Question 2.2

Mathematical induction is a method for proving the truth of propositions about the properties of integers. It is a two-step procedure whose validity rests on this syllogism of deductive logic:

The proposition is true for $n = n_0$.
If it is true for $n = n_i$, then it is true for $n = n_i + 1$, where $n_i \geq n_0$.
Therefore it is true for all $n \geq n_0$.

The first step corresponds to the first premise in the syllogism: we show the truth of the proposition for some base value, n_0, of the index, n. This base value is commonly 0 or 1, but sometimes a larger value is appropriate (see proposition (ii) below). In the second step, corresponding to the second premise, we show the truth of the logical recurrence over two successive values of the index. The conclusion then follows necessarily as a matter of valid deductive logic. (By '*valid* deductive logic', we mean that it is impossible for the conclusion to be false if the premises are true.)

You could try out your understanding of mathematical induction by proving each of the following two propositions:

i) $1^2 + 2^2 + 3^2 + \ldots + n^2 = n(n + 1)(2n + 1)/6$ for $n \geq 1$;
ii) $2^n > n^2$ for $n \geq 5$.

Despite its name, mathematical induction clearly does not use inductive logic – which is (as explained earlier in this chapter) the logic of statistical inference. It is a purely mathematical technique, and not at all statistical. To put an end to any confusion, it would be better if it were always called by its alternative name, *proof by recurrence* – as proposed many years ago by Tobias Dantzig in Appendix 12 of his book (Dantzig, 1947).

Question 2.3

Having discovered, for a few successive values of n, that the expression $n^2 + n + 41$ is prime, you might conjecture (wildly) that $n^2 + n + 41$ is prime for *all* values of n. As accumulating evidence keeps being entirely supportive, you might want to take up your initial conjecture as a plausible hypothesis. In technical language, we say 'on the accumulating evidence, this hypothesis is inductively strong'. The more values of n there are that yield a prime value, the greater the inductive strength of the hypothesis, and the more likely it is that the hypothesis is true.

In a probabilistic context, this is the line of reasoning that a statistician follows in testing a hypothesis. Because the statistician cannot neglect the influence of chance, he or she can never be certain that a hypothesis is true (i.e. cannot *prove* it is true), no matter how inductively strong it is.

However, our problem here is not stated in a probabilistic context, so it is a mathematical problem, not a statistical one. And we want to know not whether it is *likely* that the hypothesis is true (for all values of n), but whether it *is* true.

An obvious way to try to decide this is to go on evaluating the expression for ever larger values of n, to see whether, eventually, a composite (i.e. non-prime) value is generated. If such a composite value does turn up, the hypothesis is clearly false. This would be 'proof by contradiction' (or, alternatively, 'proof by counterexample'). If you proceed in this way, you will discover that for $n = 40$, the expression has the value 1681, which is 41^2. Thus, the hypothesis is actually false.

An alternative approach is to notice that substituting $n = 41$ will result in a composite number, since all three terms will have 41 as a factor.

Note: The expression $n^2 + n + 41$ as a consecutive generator of primes for n up to 39 was first given by the great Swiss mathematician, Leonhard Euler, in 1772. For more on this and other expressions that consecutively generate primes, see Beiler (1966), pages 219–221.

Question 2.4

Noting how extraordinary it seems for Colonel Openshaw to have swapped an appealing lifestyle in America for an unappealing one in England, Sherlock Holmes forms a 'working' (i.e. tentative) hypothesis that Openshaw left America in fear of someone or something. Similarly, a statistician, encountering some surprising (i.e. out of the ordinary) numerical facts, formulates a tentative hypothesis that there is some good reason for these facts.

Holmes seeks to test his hypothesis by seeing how plausible it is. One way to do this is to try to establish whether there was, indeed, someone or something *consistently* threatening Openshaw. His attempt to identify the writer (or writers) of some threatening letters is his first step in this process. Similarly, a statistician seeks to test his or her hypothesis by collecting some more data, and checking to see whether they continue to be surprising (i.e. whether they are *consistently* 'out of the ordinary', and not what might have been thrown up on a random occasion by chance influences alone).

Both Holmes and the statistician are working in conditions of uncertainty. No conclusion (i.e. inference) is sure; everything is probabilistic – as Holmes, himself, acknowledges.

Deductive logic provides systematic rules for drawing conclusions that *must*, on the basis of true premises, be true. However, in conditions of uncertainty, deductive logic is inapplicable. Instead, it is inductive logic that provides systematic ways of drawing conclusions that are *highly likely* to be true. Therefore, under uncertainty, it is inappropriate to speak of 'deducing' a conclusion from the evidence. The best that can be done is to 'induce' a conclusion from the evidence.

Interestingly, the verb 'to infer' (meaning 'to conclude') is used in both deductive and inductive logic. Thus, there is deductive inference, and there is inductive inference – though, of course, they operate according to different rules.

Throughout all the Sherlock Holmes stories, Conan Doyle makes Holmes say that he excels at observation and deduction but, in every instance, it is observation and induction that Holmes practises. Clearly, Conan Doyle was neither a statistician nor a logician!

Question 2.5

Because it is so counterintuitive, many people find it surprising to learn that the value of the *constant* π can be approximated by the results of a *random* experiment. The probabilistic result underlying this experiment was established by the French polymath and nobleman Georges Louis Leclerc, Comte de Buffon (1707–1788). The source is his *Essai d'Arithmétique Morale*, published in Paris in 1777.

Buffon posed and solved the following problem: if a needle of length m units is tossed at random onto a surface ruled with parallel straight lines d units apart, with $d > m$, what is the probability that the needle will lie across a line? Buffon approached the solution geometrically, so initiating the field now known as 'geometrical probability' (in contrast to the arithmetic approach, now known as 'combinatorial probability', which had been pioneered a century earlier by Pascal, Fermat and Bernoulli).

Buffon found the probability to be $2m/\pi d$. So, if $d = m$, the probability is simply $2/\pi$. Proving this involves a relatively simple use of integral calculus. Proofs are widely published. There is one by George Reese, online at [2.3]. The larger the number of tosses, in theory, the better is the resulting approximation of π.

By exploiting the speed of a computer, a huge number of random tosses of a needle may be very quickly simulated by means of software incorporating a suitable random number generator (on random number generators, see Chapter 11). The web now offers many sites where an applet, designed for such simulations, will generate an approximation of π on request. Reese, online at [2.3], is among them. It is interesting to investigate (i) whether the accuracy of approximation depends in any systematic way on the size of *m* relative to *d*, and (ii) how many tosses are required to obtain an approximation accurate to, say, four significant digits – that is, 3.142.

Many variants of Buffon's version of the needle problem have subsequently been devised. One of the earliest was presented in 1812 by Laplace, who generalised Buffon's surface, ruled with a set of parallel lines, to a grid of rectangles, with a second set of parallels running perpendicular to the first. The approximation to π obtained in this version is derived by Arnow (1994) and, online, by Eric Weisstein at [2.4].

References

Print

Arnow, B.J. (1994). On Laplace's extension of the Buffon needle problem, *The College Mathematics Journal* **25**, 40–43.
Beiler, A. (1966). *Recreations in the Theory of Numbers*. Dover.
Dantzig, T. (1947). *Number, The Language of Science*. Allen & Unwin.

Online

[2.3] http://mste.illinois.edu/activity/buffon/
[2.4] http://mathworld.wolfram.com/Buffon-LaplaceNeedleProblem.html

Answers – Chapter 3

Question 3.1

a) The 'gambler's fallacy' is the popular name given to the insistent belief that the chance of an event happening increases or decreases depending upon recent occurrences, when rationally it is known that the probability of that event occurring is fixed, and that successive occurrences of the event are independent. Our example of this fallacy relates to the long-term non-appearance of the lottery number 53. The gambler's fallacy can also be found operating in symmetric fashion: had a continuous run of the number 53 turned up in the lottery, there would be gamblers who were ever surer that a number other than 53 was 'due' to appear. This subjective conviction, regarding the 'dueness' of particular outcomes in a long sequence of random occurrences (as in a gambling game played continuously), arises from faith in an *empirically unsound* principle dubbed 'the law of averages'.

Now, it is true that the observed relative frequencies of all the possible outcomes of a random experiment (e.g. a lottery) will approximate well the underlying theoretical probabilities over a long run of plays, so long as the experiment is conducted fairly (that is, in line with the rules according to which the theoretical probabilities are determined). This *empirically sound* proposition is known by statisticians as the 'law of large numbers'.

However, believers in the law of averages give little thought to the meaning of 'a long run of plays'. They observe the outcomes of what amount to short runs, but still expect to see what the law of large numbers predicts. Thus, if one outcome (say) is persistently under-represented early in the observed run of plays, they think there must be some sort of 'force' at work, compelling that outcome to appear disproportionately more often in later plays. In fact, there is no such force, and consequently no law of averages. It is worth adding that, philosophically speaking, 'a long run of plays' can never be equated with any specific number: it is simply a run long enough for the observed relative frequencies of outcomes to approximate well the underlying theoretical probabilities!

All this amounts to saying that those who trust in the gambler's fallacy and, by implication, in the law of averages, are actually believers in a 'law of small numbers'. Putting it slightly differently, belief in the law of small numbers leads people to exaggerate the degree to which any particular small sample resembles the population from which it was drawn. Empirically unsound as this 'law' may be, it has a powerful grip on the psyche of many people, and governs their behaviour in the face of uncertainty in many situations. This remarkable phenomenon was first studied closely by Amos Tversky and Daniel Kahneman. Their seminal paper (Tversky and Kahneman, 1971) is titled 'Belief in the law of small numbers'. They gave a broader exposition of their ideas three years later in Tversky and Kahneman (1974). Their work has generated much subsequent research on how people actually behave in the face of uncertainty.

For further interesting reading on the gambler's fallacy and the law of small numbers, we suggest: (i) the Wikipedia entries for these two subjects, online at [3.9] – the entry for 'law of small numbers' points out usefully that there are now at least three different senses in which the term is used in mathematics and statistics; and (ii) Bruce (2002), a Sherlock Holmes pastiche in which the gambler's fallacy comes under scrutiny.

b) The probability that a specific number, such as 53, does not come up in a single drawing in the Venice lottery is $^{89}C_5/^{90}C_5 = 17/18$. Thus, the probability that it does not come up in 152 (independent) draws is $(17/18)^{152} = 1.7 \times 10^{-4}$, or about 1 chance in 6000. The probability that 53 does not come up 152 times and then comes up on the 153rd draw is $(17/18)^{152} \times 1/18 = 9.4 \times 10^{-6}$, or just less than one chance in 100,000 – obviously, a very unlikely event.

Of course, there are 90 numbers available for selection, and any of them could have been the number that didn't turn up 152 times and then turned up on the 153rd draw. Furthermore, while 152 'misses' in a row is surprising, so too would be 130, or even 100. The chance that at least one number is missing in a long sequence of draws would have a much higher and therefore less surprising probability, yet it would have been likely to set off a similar betting frenzy among a public that were not accustomed to weighing probabilities.

In fact, there have been several long 'losing streaks' in the 400-plus year history of Italian lotteries. The longest may have been 201 draws in 1941, during which the number 8 did not turn up, raising suspicions that Mussolini was fiddling the results to help finance Italy's entry into World War II (see online at [3.10]).

Question 3.2

The first question is: how representative of the nations' sleeping behaviours are the sleeping behaviours in the corresponding samples? Given that the survey was carried out via the internet, it will have missed people who lacked access to this medium. Perhaps the people who *were* reached were more likely to be interested in the internet than the rest of the population, and it is web surfing that is keeping them up late at night!

The second question is: is the sample randomly selected from its population? If it is not, then generalising on the basis of the sample information is likely to produce misleading results. In particular, the conventional statistical confidence interval formula would be unreliable

Bearing these issues in mind, but proceeding to the results nevertheless, we might next ask: how many respondents were there in each country? All we are told is that there were about 14,000 respondents from the 28 countries. Because these are countries with vastly different populations, a survey of this kind should ideally be conducted by stratifying the sample according to relative population sizes. In the absence of specific information, let us simply suppose that 500 people were sampled in each country. Assuming (rather doubtfully) that sampling was random, let's construct an approximate 99%

confidence interval around each of the sample proportions ($\hat{p}$) given, using the formula $\hat{p} \pm 2.58\sqrt{\hat{p}(1-\hat{p})/500}$. On this basis, the population proportion for Portugal could be expected to lie in the interval (0.70, 0.80), for Taiwan in (0.64, 0.74), for Korea in (0.63, 0.73), for Hong Kong in (0.60, 0.72), and for Spain in (0.59, 0.71). The overlapping intervals for these five countries suggest that national differences in the sample proportions of those who go to bed after midnight may be due to chance variation alone.

Though we have done these calculations under some doubtful assumptions, they nevertheless reveal two major weaknesses of all league tables that are based on sample data. These are, firstly, that if we look only at the column of rankings, we are ignoring valuable information – namely, the actual differences between the scores of the ranked entities. Secondly, if the scores are made available (and sometimes, regrettably, they are not!), it is often the case that some differences are so small as to be statistically insignificant. What this means is that the ranking is actually much fuzzier, both than the ranking scores indicate, and than the authors of the ranking may want to imply. (There is more about deficiencies of league tables in CHAPTER 8.)

Question 3.3

For making valid statistical inferences, isn't it more important to have a representative sample than a random sample? The short answer is no, because randomness is *indispensable* for valid statistical inference. A sample, once selected, *may turn out* to be representative but, first and foremost, it *must* be randomly selected. To see why randomness is so fundamental, let us look at a specific context.

Suppose the manager of a dental practice with 1500 adult patients wants to survey these patients on their oral hygiene activities (brushing, flossing, etc.). She settles on a sample size of 200 adult patients, and her initial thought is to survey the first 200 that come into the practice. After a little reflection, she realises that oral hygiene practices very likely differ with age, and so decides to survey patients in four age groups: 16–24, 25–39, 40–59, 60 and over. She also recognises that she will not get reliable results if she simply approaches the first 200 adults who arrive. Why? Because it is possible that a particular age group will be very under-represented in her sample.

So, to guide her data collection systematically, she decides to use a representative sample – that is, one where the sample size in each age group is the same proportion of 200 as the number of adult patients in each age group is of 1500. From patient records, she finds the following percentages of patients in the respective age groups: 16%, 37%, 35% and 12%. Thus, she seeks samples of size 32, 74, 70 and 24 in the corresponding age groups.

She asks the age of each patient on arrival, and then administers the questionnaire. Proceeding this way, she will very likely need to interview more than 200 patients, because they will not necessarily arrive in numbers that immediately fill her quota for each age group. In due course, she produces a statistical report showing what proportion of patients in each sampled age group brush twice a day, and so on. She interprets these statistics as valid point estimates of the corresponding proportions for all adult patients in the practice, and constructs confidence intervals around these point estimates. Unfortunately, as already mentioned, these various estimates are *not* valid, because they are based on data from non-random samples.

It may seem paradoxical that sample randomness is prioritised over sample representativeness in statistical inference. There are two reasons for prioritising randomness – one is theoretical, the other practical. The theoretical reason is that all the optimal properties of confidence intervals (and other statistical inferences) are based on probability distributions valid only for random samples. The practical reason is that any systematic procedure for sample selection risks inadvertently including in the sample (or excluding from it) particular data values. Each of these actions may bias the conclusions.

Consider the manager's intention to construct her representative sample only from among those who come into the practice. Her database is likely to be deficient in data from those whose dental health is either very good (because they brush conscientiously) or very bad (because of long-term neglect). Lacking data from the 'extremes', her statistical analyses may well lead to biased conclusions.

A statistically appropriate procedure for the manager to follow in this context – were she to adopt *simple random* sampling – would be to use a table of random numbers to draw a random sample of 200 from all 1500 patient record numbers. However, she has decided to use *stratified random* sampling, where the strata are the four age groups. Then it is quite in order for her to use sample sizes of 32, 74, 70 and 24 in the respective strata (as explained above), but she must draw these four separate samples randomly from the corresponding four lists of all 1500 patients classified by age group.

The multiple meanings of the expression a 'representative sample' were investigated by two eminent US statisticians, William Kruskal and Frederick Mosteller. So extensive was their exploration, it was reported in no fewer than four articles (see Kruskal and Mosteller, 1979–80). If you enjoy discovering subtlety and richness in language, all four are worth your attention. We recommend, in particular, the third of these articles, which focuses on statistical contexts. The authors discern nine senses in which statisticians have used a 'representative sample', none of which is identical to the formal definition of a 'random sample'. It becomes clear that to deliberately confuse a representative sample with a random sample can be an effective ploy to mislead the public on the validity of statistical inferences. Regrettably, this happens often.

We look at more such ploys, and other inappropriate uses of statistical techniques, in CHAPTERS 8 and 9.

Question 3.4

Apart from the number of class intervals, two other features of a histogram influence its shape, namely the choice of widths for the class intervals, and the specific values that are imputed as the absent boundaries of the first and last class intervals, if they are open-ended.

Several decades ago, statistics textbooks used to mention Sturges' Rule for defining the most practical number of class intervals for grouping data. Sturges' Rule (which assumes that class intervals have equal width) says: choose the number of class intervals to be the integer nearest in value to $1 + 3.3 \log_{10} n$, where n is the number of observations to be grouped. This rule of thumb seeks a compromise between choosing very few class intervals (when too much information on the variability of the data will be lost in the grouping) and very many class intervals (when too little data summarisation will be achieved). The reasoning behind this formula can be followed in the original paper by Herbert Sturges (1926).

Question 3.5

Can we estimate the mean of a (finite) population from values in a nominated sample? Yes, but the task in the context of this question is not straightforward, and the result is not obviously dependable.

Firstly, we need a way of finding the mean of a sample of data presented not individually, but already grouped in a histogram. This can be done by making a thin card cut-out of the histogram, and finding where it balances on a laterally-movable knife edge set at right angles to the horizontal axis. In this way, we estimate the sample mean value to be close to 26 years.

Secondly, we know nothing about the way the sample was selected. The ages given are recorded for only around one-third of the convicts (see Cobley, 1970). There is no evidence of any deliberate bias in the selection of those whose ages were recorded, but there is also no certainty that the recorded ages comprise a random sample. So, while it is perfectly possible to use these data to estimate the mean age of all the convicts on the First Fleet, the estimate will be unreliable unless the sample is indeed random.

Note: in the first two paragraphs of this answer, we turned up two uses of the word 'estimate'. Initially, we used 'estimate' in the technical statistical sense of 'the inductive step of passing from a sample value to a population value'. Thereafter, 'estimate' appeared in its more common meaning of 'approximate measurement or calculation'. You will find the same dual usage in the wording of QUESTION 18.3.

References

Print

Bruce, C. (2002). The case of the gambling nobleman, in his *Conned Again, Watson! Cautionary Tales of Logic, Maths, and Probability*. Vintage.

Cobley, J. (1970). *The Crimes of the First Fleet Convicts*. Angus & Robertson.

Kruskal, W. and Mosteller, F. (1979–80). Representative sampling. *International Statistical Review*: part I, **47**(1), 13–24; part II, **47**(2), 111–127; part III, **47**(3), 245–265; part IV, **48**(2), 169–195.

Sturges, H.A. (1926). The choice of a class interval. *Journal of the American Statistical Association* **21**, 65–66.

Tversky, A. and Kahneman, D. (1971). Belief in the law of small numbers. *Psychological Bulletin* **76**, 105–110.

Tversky, A. and Kahneman D. (1974). Judgment under uncertainty: heuristics and biases, *Science*, **185**, 1124–1131.

Online

[3.9] http://en.wikipedia.org/wiki/Law_of_small_numbers and
https://en.wikipedia.org/wiki/Gambler's_fallacy

[3.10] http://www.guardian.co.uk/international/story/0,,1410524,00.html

Answers – Chapter 4

Question 4.1

The writer is Girolamo Cardano (1501–1576) in his *Liber de Ludo Aleae*, published posthumously in 1663. The passage from Cardano's work is quoted by Anders Hald in his book, *A History of Probability and Statistics and Their Applications before 1750*. Cardano was well qualified to comment, since he admitted to being an inveterate gambler: 'During many years – for more than forty years at the chess boards and twenty-five years of gambling – I have played not off and on but, as I am ashamed to say, every day'. (Hald, 1990, page 38). Attitudes today to the treatment of compulsive gambling seem to have caught up with his ideas!

Cardano was also an able mathematician, who wrote several treatises summarising the mathematical knowledge of his day. His life and contributions to the early understanding of probability are interestingly reviewed in chapters 5 and 6 of David (1962).

Question 4.2

The play is *Rosencrantz and Guildenstern are Dead* by Tom Stoppard. As the play opens, the coin tossing is in progress and already a run of 69 heads has been obtained. Stoppard writes in the stage notes, 'the run of "heads" is impossible', which is clearly false! As the audience watches, the run of heads continues to 100. Guildenstern comments: 'A weaker man might be moved to re-examine his faith, if in nothing else at least in the law of probability.' Behind the characters' further philosophising, based on half-remembered and only partially accurate ideas about probability, is the notion that this highly unusual event (the long string of heads) suggests something is wrong with the world. There is also a film of the play, Stoppard (1990).

Question 4.3

A census was commanded at Christmas 1085 by William the Conqueror (1028?–1087), who had invaded England from Normandy in 1066. The census took a little over a year to complete but the written record, known as the Domesday Book, was never completed in full detail. Probably on account of changed priorities following the death of William in 1087, some of the later entries were just summaries of the data collected.

Today, the Domesday Book is held in the UK National Archives. There are, in fact, two books – Great Domesday and Little Domesday – totalling about 1000 parchment pages. These books record, in Latin, the ownership and other details of all land in over 13,000 places in England and Wales.

An ingenious reanalysis of some of the rich data in the Domesday Books, using modern econometric methods, is described in McDonald and Snooks (1985), subsequently expanded by the authors into a most unusual book (McDonald and Snooks, 1986).

Question 4.4

The article was by Edgar Anderson (1935), entitled 'The irises of the Gaspé Peninsula'. The Gaspé Peninsula (in French, Péninsule Gaspésienne) is in Quebec, Canada, south of the St. Lawrence River. The data consist of length and width measurements of sepals and petals in samples of 50 of each of three species of iris (*Iris setosa*, *Iris virginica* and *Iris versicolor*). This set of data was used soon after by R.A. Fisher (1936) to illustrate his newly-developed technique of discriminant analysis. Fisher used only the measurements on *Iris setosa* and *Iris versicolor*, but the whole data set is widely available on the web – for instance, at the Wikipedia site [4.13]. Fisher's 1936 paper can be found in his Collected Papers, online at [4.14].

Question 4.5

The *Loterie de France* was first proposed by Giacomo Casanova (1725–1798) to Louis XV, and ran from 1758 to 1836. In the earliest version, players bet on a single number (an *extrait*), a pair of numbers (an *ambe*) or a triple of numbers (a *terne*). Five winning numbers were then officially drawn from the numbers 1 to 90. The gambler won if all his or her selected numbers were among the winning numbers. An engaging 2003 article by Stephen Stigler about the history of the *Loterie* is online at [4.15].

In contemporary games of lotto, winners are paid from a 'pari-mutuel pool', that is, a certain proportion of the money collected from the sale of tickets. In the Australian *OZ Lotto*, for example, this dividend pool comprises 56.5% of the total collected from ticket sales. In Casanova's *Loterie*, the prizes were backed by the French Government, and Louis XV ran the risk of losses of up to a hundred million francs – although, of course, the odds were very much against this happening.

References

Print

Anderson, E. (1935). The irises of the Gaspé Peninsula. *Bulletin of the American Iris Society* **59**, 2–5.

David, F.N. (1962). *Games, Gods and Gambling – A History of Probability and Statistical Ideas*. Griffin.

Fisher, R.A. (1936). The use of multiple measurements in taxonomic problems. *Annals of Eugenics* **7**, 179–188.

Hald, A. (1990). *A History of Probability and Statistics and Their Applications before 1750*. Wiley.

McDonald, J. and Snooks, G.D. (1985). Statistical analysis of Domesday Book (1086). *Journal of the Royal Statistical Society, Series A* **148**, 147–160.

McDonald, J. and Snooks, G.D. (1986). *Domesday Economy: a New Approach to Anglo-Norman history*. Oxford University Press.

Stoppard, T. (director and writer) (1990). *Rosencrantz and Guildenstern are Dead*. Film, 117 minutes. Production: Brandenberg and WNET Channel 13, New York.

Online

[4.13] http://en.wikipedia.org/wiki/Iris_flower_data_set
[4.14] https://digital.library.adelaide.edu.au/dspace/bitstream/
        2440/15227/1/138.pdf
[4.15] http://president.uchicago.edu/page/
        2003-nora-and-edward-ryerson-lecture

Answers – Chapter 5

Question 5.1

As everyone has experienced when feeling unwell, sometimes this feeling is accompanied by a fever – a rise in bodily temperature above one's 'normal' temperature. So at ease are we with how it feels to be at 'normal' temperature, we register our bodily temperature only when it rises (or falls) abnormally – hence, the otherwise illogical question, 'does she have a temperature?'

In fact, however, there is no single 'normal' temperature for a well person. There can be any number of 'normal' temperatures, depending on the circumstances! Just as an indication: a well person's temperature varies with the site of measurement (in the armpit, in the mouth, etc.), with the person's emotional state, with the time of day, and with the ambient temperature and humidity. Even if the temperatures of many individuals in identical circumstances are averaged, the averages also differ systematically *across* different circumstances, by as much as 1.5 Celsius degrees. It follows that 37.0°C (or its Fahrenheit equivalent, 98.6°F) is no more than a conventional demarcator between what is normal and what is abnormal. It would seem to be more meaningful to have international agreement on a range for what is called normal: 35.8–37.2°C appears plausible to us.

Question 5.2

Population statistics are changed from time to time on the cited Wikipedia page. These changes presumably reflect official estimates of population change – estimates which can be no more accurate than the statistical models and methods used to generate them. Area statistics are United Nations evaluations, presumably from satellite mapping. Population density is the ratio of population to area.

Populations and areas vary enormously among the world's countries, and the same population density (ratio) can arise from very diverse numerators and denominators. For example, Chile and Brazil have closely similar population densities (22 and 24 persons per square kilometre, respectively) but their populations are vastly different, as are their areas. Density comparisons are likely to be more meaningful between countries with similar populations, or with similar areas. Thus, the Netherlands and Chile are similar in population size, but have very different areas. Again, Brazil and Australia have similar areas, but very different populations. It is also very relevant to consider countries' topographies in interpreting a comparison of population densities. While Canada and the USA have similar areas, much more of Canada, than of the USA, is inhospitable to human habitation.

Question 5.3

The letter contains an array of muddled statistical and non-statistical ideas. Taking the writer's points in the order he makes them, we find the following problems.

First, an A-to-E grading system is, by itself, neither more nor less logical than a pass/fail grading system. The former is generally conceived as an assessment on a *relative* standard – that is, relative to what others at the same level of education know and understand; while the latter is an assessment on an *absolute* standard – that is, as judged against some mandated minimally acceptable level of understanding. In any case, an A-to-E grading system is not more logical because it has a five letter range, rather than (say) a seven letter range; the five letter range is simply more familiar.

Second, the sweeping statement that 'most people are "average" or C on any measure (as is shown in the bell curve)' is not necessarily correct. If: (i) the measure were continuous; (ii) values on the measure were approximately normally distributed; and (iii) the grades A to E corresponded to equal intervals on the measure, it would be correct to say that more people would be in category C than in any other category – though, even then, this may still not be 'most people'!

Third, the statement that an A-to-E ranking gives below-average students 'some hope of improving' may be well-intentioned, but logically it is no more meaningful than to say that students who have received a 'fail' grade in a pass/fail framework can have some hope of improving.

Fourth is the concern that a letter grade received in one school may not be equivalent to the same letter grade received in another school. This depends, of course, on the criteria for assigning the grades. If the criteria were state- or nationwide, the concern would be resolved.

Finally, there is the writer's confident presumption that attending an independent school allows students to earn letter grades higher than they would if attending a government school. This view ignores the fact that student learning is dependent, in part, on a whole range of factors (such as interest, commitment, family background, standard of living) that are not necessarily related to the auspices of the school.

Question 5.4

The truth about the quotation widely attributed to Wells is investigated in Tankard (1979). Tankard shows how Wells' original words became transformed, as several later writers – perhaps a little impatient at Wells' prolix writing style – progressively misquoted him, presumably so as to sharpen the central point that *they* wanted to make.

You can find Wells' original words on this theme by searching for the relevant occurrence of the phrase 'read and write' in any of the online editions of *Mankind in the Making* (for example, that of Project Gutenberg, online at [5.9]). Tankard, of course, gives these words as well.

This is what Wells actually wrote (in his chapter 6, entitled *Schooling*): 'The great body of physical science, a great deal of the essential fact of financial science, and endless social and political problems are only accessible and only thinkable to those who have had a sound training in mathematical analysis, and the time may not be very remote when it will be understood that for complete initiation as an efficient citizen of one of the new great complex world-wide states that are now developing, it is as necessary to be able to compute, to think in averages and maxima and minima, as it is now to be able to read and write.'

Note particularly that, while Wells mentions 'averages', the term 'statistical thinking' does not appear at all!

Question 5.5

There are two forms of frequency distribution in which few observed values are likely to be found at or near the average: bimodal distributions and U-shaped distributions. Bimodal distributions are common in real-world contexts. For example, the frequency distribution of the age of pedestrians killed in road traffic accidents is strongly bimodal, with the youngest not yet understanding that moving vehicles are dangerous, and the oldest unable to get out of the way in time, despite understanding this all too well! Genuinely U-shaped distributions are more unusual in practice. An example is the distribution of the percentage of a city's residents in a particular age group requiring full-time care.

Why *does* an average alone provide only a minimally meaningful picture of a frequency distribution? Because it cannot provide any information about the shape of the distribution. One source of information about shape is a measure of how spread out the values are across the distribution. Oddly, such a measure is rarely provided when data, implicitly from a frequency distribution, are presented in the popular media. CHAPTER 6 agitates for change in this state of affairs.

References

Print

Tankard, J.W. (1979). The H.G. Wells quote on statistics: a question of accuracy. *Historia Mathematica* **6**, 30–33.

Online

[5.9] http://www.gutenberg.org/etext/7058

Answers – Chapter 6

Question 6.1

By direct substitution, $A = \dfrac{2}{9}[X_1^2 + X_2^2 + X_3^2 - X_1X_2 - X_1X_3 - X_2X_3]$ *and*

$B = \dfrac{2}{3}[X_1^2 + X_2^2 + X_3^2 - X_1X_2 - X_1X_3 - X_2X_3]$. So here, B = 3A. In general, for N observations from a discrete distribution, $B = [2N/(N-1)]A$. Thus, as N increases, B approaches 2A from above. It follows that, for values of N usual in practice, the square of the intuitive measure of variability is systematically just about twice the square of the standard deviation. In symbols (where M is the intuitive measure, and σ the standard deviation), $M \approx \sigma\sqrt{2}$. When the data are from a continuous distribution with an infinite range, it can be shown that the relationship is exact, i.e. $M = \sigma\sqrt{2}$.

Historical note: our intuitive measure of variability was recognised by Francis Edgeworth in 1885 as a logically meaningful measure of dispersion. He called it the 'modulus' (adopting a term that had been introduced by de Moivre in 1738, but with a slightly varied definition), and it continued in use until the end of the 19th century. The concept of the standard deviation was known in Edgeworth's time – not by that name (which was first used in print by Karl Pearson in 1894) but, rather, by the names 'root mean square error' or, less intelligibly, 'error of mean square'.

These two measures of dispersion, distinct but closely related, were among several others in circulation at that time. In particular, a measure called the 'probable error' was introduced by the mathematician Friedrich Bessel in 1815, and became more prevalent than the modulus. It was defined for symmetric distributions (notably the normal distribution), and it is identical to what we today call the semi-interquartile range. Thus, for the normal distribution, in symbols (where P is the probable error), $P = 0.67\sigma$.

Such was Pearson's standing in the field of statistics that, soon after 1900, the standard deviation (both the concept and the name) swept away the modulus and the probable error. It was R.A. Fisher who, in 1918, coined the term 'variance' as a shorthand for 'square of the standard deviation.'

Question 6.2

No, the variance may decrease, remain the same, or increase. Deletion of the largest value of X always reduces $\Sigma(X_i - \mu)^2$, but N also reduces by 1, so the ratio will not necessarily fall.
Example:

a) for the set [1, 13, 13, 13, 16] the variance is 27.36, while for the set [1, 13, 13, 13] the variance is 27.00, so the variance decreases;
b) for the set [1, 13, 13, 13, 15.81], the variance is 27.00, so deleting the largest value leaves the variance unchanged;
c) for the set [1, 13, 13, 13, 15] the variance is 25.60, so deleting the largest value increases the variance.

Question 6.3

If the standard errors had been reported, we could construct a confidence interval for the actual mean floor area of homes in each country. We could also test for significant pairwise differences in mean floor area between countries.

If we knew, further, (i) whether the definition of a 'home' was the same in each country, and (ii) what proportion of homes in each country had 1, 2, 3, or more bedrooms, we would gain insight on the extent to which cross-country comparisons were actually comparing like with like.

Question 6.4

The sampling variance of the median of n values, drawn with replacement from a normal distribution with mean μ and variance σ^2, is *approximately* $(\pi/2)(\sigma^2/n)$. Details of this result and its technically-advanced derivation are given in Stuart and Ord (1994), chapters 10.10 and 14.6.

Question 6.5

We are considering two alternative estimators of the population variance, σ^2: s^2 denotes the sample variance, and $\hat{\sigma}^2$ the sample analogue of the population variance. The information given shows that s^2 is unbiased for σ^2, while $\hat{\sigma}^2$ is biased. Perhaps that is why introductory textbooks of statistics prefer the counterintuitive s^2 over the more intuitively appealing $\hat{\sigma}^2$ – starting, indeed, from the early chapter on descriptive statistics?

But there is more to discover. How do s^2 and $\hat{\sigma}^2$ compare on efficiency of estimation? Comparing the given variances of these estimators in the normal case, it is easy to show that $\text{var}(\hat{\sigma}^2) < \text{var}(s^2)$. In other words, $\hat{\sigma}^2$, though a biased estimator, is more efficient than s^2.

When two estimators of the same parameter show themselves alternately superior, in this way, on the criteria of unbiasedness and efficiency, it is constructive to compare them on their mean square error (MSE). The MSE of an estimator is defined as its variance plus the square of its bias. In the normally distributed case, $\text{MSE}(s^2) = 2\sigma^4/(n-1) + 0^2 = 2\sigma^4/(n-1)$. And $\text{MSE}(\hat{\sigma}^2) = 2\sigma^4(n-1)/n^2 + [-\sigma^2/n]^2 = \sigma^4(2n-1)/n^2$. It can readily be shown that $\text{MSE}(\hat{\sigma}^2) < \text{MSE}(s^2)$.

One may now really wonder why s^2 gets the accolade so comprehensively in introductory statistics, when $\hat{\sigma}^2$ has both greater efficiency and smaller MSE. See Sahai and Misra (1992).

There is a further complication to be faced in practice when constructing an interval estimate of a population mean. This requires an estimate of the standard error of the sample mean, $\sigma/\sqrt{n}$, and that, in turn, needs an estimate of the 'nuisance' (i.e. impeding) parameter σ. Even though s^2 is unbiased for σ^2, unfortunately neither s nor $\hat{\sigma}$ is unbiased for σ (this is a quite general situation: unbiasedness is not preserved under square root transformation). Despite the availability of several unbiased estimators of σ (see, for instance, Cureton (1968), and D'Agostino and Cureton (1973)), textbooks invariably adopt s as the estimator of σ in this context. How totally puzzling!

What is the explanation? The key to the preference for s^2 and s in these contexts is the superior tractability of the statistical distribution theory for any consequent inference – involving an interval estimator or test statistic – regarding the population mean. This is simply one example of a phenomenon familiar to experienced statisticians – an arguably suboptimal *theoretical* choice of 'nuisance' parameter estimator along the way may be tolerated, because it ultimately produces a more workable *practical* inferential procedure for the parameter of principal interest.

References

Print

Cureton, E.E. (1968). Unbiased estimation of the standard deviation, *The American Statistician*, **22**(1), 22.

D'Agostino, R. and Cureton, E.E. (1973). A class of simple linear estimators of the standard deviation of the normal distribution. *Journal of the American Statistical Association* **68**, 207–210.

Sahai, H. and Misra, S. (1992). Definitions of sample variance: some teaching problems to be overcome. *The Statistician* **41**, 55–64.

Stuart, A. and Ord, J.K. (1994). *Kendall's Advanced Theory of Statistics*, volume 1, 6th edition. Arnold.

Answers – Chapter 7

Question 7.1

The simple AM index, calculated as an *average of price ratios* shows spuriously, as we saw in this chapter's Overview, that the general price level is rising. The simple AM index, calculated as a *ratio of price averages*, is $[(4 + 3)/2]/$ $[(2 + 6)/2] = 0.875$. This indicates, spuriously, that the general price level is falling. The simple GM index, calculated as a *ratio of price averages*, is $\left(\sqrt{4 \times 3}\right) / \left(\sqrt{2 \times 6}\right) = 1$. This indicates (like the simple GM *average of price ratios*) no change in average price level.

For constructing a meaningful price index, *neither* of the alternative measures involving the AM is appropriate. As for the alternative GM-constructed indexes, in every situation they both always give the same result.

You may have noticed that, by choosing self-servingly among these formulae, it is possible to demonstrate any of the following: average price up, average price down, or no change. As with many other statistical tools (see, especially, CHAPTERS 8 and 9), there is scope for deliberate misuse of price index formulae.

Question 7.2

The intuition underlying this question is quite a simple one: if there are two goods and an increase in one good's price is accompanied by an equal decrease in the other good's price, then there has been no change in the average price of the two goods. In the split second of deciding whether to trust this intuition, most people would probably think of a fixed *absolute* (dollar amount of) increase or decrease in price. Their past experience with the average everyone knows – the arithmetic mean – would then immediately signal that this intuition could be trusted.

However, the context of this question is actually that of a fixed *proportional* increase or decrease in price. Even if they notice the different context, it is still highly likely that most people will unhesitatingly go on trusting the same intuition. So the fundamental question here is, 'is this unhesitating carry-over of trust justified?' The answer is 'yes'; but there is more to this answer, as the following discussion shows.

Suppose there is a fixed proportional increase, k, in the price of one good between two points in time (say, from price p_1 to price kp_1), and the same proportional decrease, written $(1/k)$, in the price of the other good (say, from price p_2 to $(1/k)p_2$). The ratios of prices are kp_1/p_1 and $(1/k)p_2/p_2$. Since each number has a unique logarithm, we may equally well work with $\log[kp_1/p_1]$ and $\log[(1/k)p_2/p_2]$.

The arithmetic mean is already accepted as a trustworthy way of finding the average price. Thus, if we find the arithmetic mean of the two log price ratios and then the antilog of this arithmetic mean, we expect also to have a trustworthy average price. It is highly unlikely, of course, that a non-statistician's intuitive thinking would actually proceed along this chain of thought!

Thus, while the carry-over of trust in intuition about an average is justi-fied, the intuition is not, in this context, true about the average that most people have in mind – that is, the arithmetic mean (AM) of price changes. Instead, the reasoning we have just spelled out has generated the geometric mean (GM) of price changes. Recall that the logarithm of the GM of a set of numbers is the AM of the logarithms of those numbers. This example illus-trates how one's intuition may not always be a completely trustworthy guide in statistical work.

There are, indeed, lots of counterintuitive results in both statistical theory and probability theory. The Central Limit Theorem (see CHAPTER 12) is a prime example. Paradoxes are, by definition, another source, and there are many examples in CHAPTERS 10 and 11. In these examples, most people's intuitions turn out to be *entirely* wrong

What makes the context of the present question unusual is that, while most people's intuition about the in-principle trustworthiness of an average is right, the specific average that turns out to be trustworthy is unforeseen.

Should *statisticians* trust their intuition? Our view is that it depends on the scientific and personal attributes of the statistician. Someone who is already deeply knowledgeable in the subject and its intellectual history, and who is by nature cautious about making too-hasty presumptions, may be well served by his or her intuition. It is now well known, for instance, that R.A. Fisher's pioneering work on the theory of experimental design grew out of his intuitive reaction to a real-life event, as we invite you to discover in QUESTION 19.1.

Question 7.3

A weighted GM of the n values X_i, each value having the weight w_i, $(i = 1, 2 \dots n)$, has the general form:

$$GM = \sqrt[\Sigma w_i]{X_1^{w_1} X_2^{w_2} \ldots X_n^{w_n}} = \left(\prod_1^n X_i^{w_i} \right)^{\frac{1}{\Sigma w_i}} = \prod_1^n X_i^{\frac{w_i}{\Sigma w_i}}$$

Thus, a weighted GM of the n values p_{0i}, each value having the weight q_{0i}, is:

$$GM = \prod_1^n p_{0i}^{\frac{q_{0i}}{\Sigma q_{0i}}}$$

And a weighted GM of the n values p_{1i}, each value having the weight q_{0i}, is:

$$GM = \prod_1^n p_{1i}^{\frac{q_{0i}}{\Sigma q_{0i}}}$$

Then the GM analogue of the Laspeyres price index is $\left. \prod_1^n p_{1i}^{\frac{q_{0i}}{\Sigma q_{0i}}} \middle/ \prod_1^n p_{0i}^{\frac{q_{0i}}{\Sigma q_{0i}}} \right.$.

Question 7.4

If you invest $48,000, at the rate of $100 per month over 40 years, and watch this sum grow by compound interest to $632,408, why is it practically misleading to declare that your capital has been multiplied about 13 times? Because in any real-world context, a given amount of money will not continue to buy the same quantity of products as time goes by. Consider this: in 1976 in Australia, the cost of a six-cylinder family car was of the order of $5,500 – but today, such a car costs around $40,000. Evidently, a dollar today buys much less car than a dollar in 1976. Not all of the price rise in cars reflects technological improvements. Some of it echoes the gradual rise, over time, in the prices of most goods we buy as consumers. This general rise in the price level over time is termed *inflation*.

To abstract from the pervasive influence of inflation on consumer prices, we compare money amounts in terms of 'fixed-year' dollars. For this, we need a consumer price index. Such an index is calculated by the national statistical office in most countries.

Let's suppose that the investment described in this question took place in Australia between 1976 and 2015. Using the Australian consumer price index for these 40 years, we can estimate the value, *in terms of 1976 dollars*, of $100 deposited monthly over 480 months. This is approximately $18,000. Similarly, we can work out the value, *in terms of 1976 dollars*, of a lump sum of $632,408 collected in 2015. That is approximately $111,000. So, *in terms of 1976 dollars*, your capital has not been multiplied about 13 times, but only about 6.2 times. The 'wonders' of compound interest are evidently not as wonderful as all that!

It will be clear now that, when it comes to comparing money amounts over time, comparisons which do not take into account the effects of inflation on the purchasing power of those amounts flagrantly neglect a fundamental principle of valid data analysis in economic statistics.

Question 7.5

The Marshall-Edgeworth (M-E) index does not confound price change and quantity change. To see why, consider the general form of a price index which is (like the M-E index) a ratio of weighted means of prices (of n products). When the mean is the AM and the weights are w_i ($i = 1, 2, \ldots n$), the equation of such an index is $[(\Sigma p_{1i} w_i)/\Sigma w_i]/[(\Sigma p_{0i}w_i)/\Sigma w_i] = \Sigma p_{1i}w_i/\Sigma p_{0i}w_i$. What is significant about this form is that *exactly the same* set of values appears as weights in the numerator and the denominator. Then, regardless of how the w_i are defined, the index value generated reflects only the variation in prices.

In the case of the M-E index, $w_i = (q_{0i} + q_{1i})/2$, and this set of weights is common to numerator and denominator. Hence, an M-E index generated value reflects only the variation in prices. What, then, is an example of a confounded index? Here is one: $\Sigma p_{1i}q_{0i}/\Sigma p_{0i}q_{1i}$.

[*Note*: it is not hard to prove that, for a fixed set of data, the value of the Marshall-Edgeworth index always falls between the values of the Laspeyres and the Paasche indexes.]

Answers – Chapter 8

Question 8.1

The poem is *The Battle of the Nile* by William McGonagall (1825–1902), the self-styled Poet and Tragedian of Dundee (in Scotland). On Chris Hunt's website [8.2], McGonagall is said to be 'widely hailed as the writer of the worst poetry in the English language'. This website presents the complete text of this and other poems, as well as an autobiography.

Question 8.2

How to Lie with Statistics is described, rather extravagantly, as 'the most widely read statistics book in the history of the world' in Steele (2005), online at [8.3]. In chapter 8 of his book, Huff explains and illustrates the logical fallacy '*post hoc ergo propter hoc*' ('following this, therefore because of this'). The name of this fallacy says it all! We may fall into fallacy if we attribute causality to the interaction of two (or more) variables simply because we observe that their values are correlated through time. Before a correlation can be (even tentatively) labelled as causal, it is necessary first to have a theory which postulates a causal mechanism (defining both its direction and its degree of stability), and then to confirm this theory in a variety of settings where the correlation is observed. (See also the answer to QUESTION 9.1.)

Quite often nowadays, one comes upon a press report of someone's claim to have discovered (or corroborated) a causal relation in some scientific field. These reports frequently publicise recent articles in the scholarly literature. At times, however, the supposedly causal relation seems so astonishing that one wonders whether this isn't just the 'post hoc' fallacy in action. Here are two examples. First, researchers from the University of Cambridge report that a longer ring finger than index finger predicts a more successful financial trader – see Coates, Gurnell and Rustichini (2009). Second, a systematic US study concludes that frequent attendance by women aged 50–79 at religious services reduces their risk of death from all causes by as much as 20%, compared with those not attending services at all – see Schnall *et al.* (2010). We leave you to decide whether these papers have succeeded in eliminating the possibility that the 'post hoc' fallacy is lurking.

Of course, there are many, and more straightforward, situations where observing people's behaviour in daily life strongly suggests that they have fallen prey to the 'post hoc' fallacy.

Question 8.3

At first sight, the statement 'obesity is costing the country $56 billion a year' might appear to refer simply to the *direct costs* of medically treating obesity, and the many other conditions (e.g. high blood pressure, diabetes and arthritis) that commonly follow.

With a little further thought, it will be apparent that there are also *indirect costs* of obesity in the community, such as the cost of providing offices and hospitals with more robust furniture to accommodate obese individuals, and the cost of running public education programmes on the dangers of obesity. Indirect costs are not necessarily actual financial outlays; some indirect costs are implicit losses to society from the value of economic output forgone by obese people on account of their obesity, such as loss of output from extended periods of sick leave. In addition, there are *opportunity costs* – that is, costs to other social welfare and medical pro-grammes (e.g. on homelessness or mental illness) imposed by the urgency of directing scarce research funds towards new treatments for obesity and its complications, rather than to those other programmes.

Without further clarification, it is impossible to know whether the state-ment on the cost of obesity to the nation – and similar statements regarding the cost of other personal and social ills that impact heavily on society – refer to all, or just to some, of the categories of costs just mentioned. Moreover, there is no way of knowing how comprehensive is the array of different costs within each category that have been evaluated in coming to the final total, nor how accurately they have been evaluated. In other words, such state-ments are only as trustworthy as the organisation under whose auspices they were produced.

Question 8.4

The series was *Yes, Prime Minister*, written by Jonathan Lynn and Antony Jay. The episode was *The Ministerial Broadcast* (series 1, episode 2), first broadcast in January 1986. In a crisply-worded exchange, Parliamentary Permanent Secretary, Sir Humphrey Appleby, reveals to the PM's Principal Private Secretary, Bernard Woolley, how to devise opinion polls to elicit diametrically opposed responses to the question, 'Would you be in favour of reintroducing National Service?' The technique is to build up to the crucial question by two alternative paths. Each path employs a sequence of ques-tions which predisposes an uninformed (or unthinking) interviewee to express exactly the view that the interviewer has 'coached' the interviewee to state. The full text of the exchange is available online at [8.4], and the clip can be watched on YouTube at [8.5].

Question 8.5

Temperature can be measured using any one of several different scales – Celsius and Fahrenheit are the most common ones – but to cor-rectly measure thermal energy (i.e. the amount of heat), degrees on the Kelvin scale are used. That scale has an absolute zero, as opposed to the Celsius and Fahrenheit scales, each of which has an arbitrary zero. Zero on the Kelvin scale represents the complete absence of thermal energy. On the Kelvin scale, a doubling of the temperature indicates a doubling of thermal energy. There is a simple connection to the Celsius scale shown by the equa-tion $°K = °C + 273.15$. So $5°C = 278.15°K$ and $10°C = 283.15°K$, representing a 1.8% increase in thermal energy.

There is more in CHAPTER 21 on alternative scales of measurement.

References

Print

Coates, J.M., Gurnell, M., and Rustichini, A. (2009). Second-to-fourth digit ratio predicts success among high-frequency financial traders. *Proceedings of the National Academy of Sciences* **106**(2), 623–628.

Schnall, E. *et al.* (2010). The relationship between religion and cardiovascular outcomes and all-cause mortality in the women's health initiative observational study. *Psychology and Health* **25**(2), 249–263.

Online

[8.2] http://www.mcgonagall-online.org.uk/index.shtml

[8.3] Steele, J.M. (2005). Darrell Huff and fifty years of 'How to Lie with Statistics'. *Statistical Science* **20**, 205–209. At https://projecteuclid.org/euclid.ss/1124891285

[8.4] http://www.yes-minister.com/ypmseas1a.htm

[8.5] https://www.youtube.com/watch?v=G0ZZJXw4MTA

Answers – Chapter 9

Question 9.1

As the wording of the question hints, this is an illustration of the '*post hoc ergo propter hoc*' fallacy described in QUESTION 8.2. Whimsical though it is, this example is useful for analysing the serious question: how does one actually decide whether an observed association is, or is not, a sign of a causal process?

Suppose an association is observed between two variables, A and B. It might be an entirely coincidental association, or there may be a causal relation between A and B, or a third variable, C, might be acting separately on A and on B at the same time.

If there were, in fact, a direct causal relation between A and B, it would not be clear without further investigation which of the following was true: (i) change in A causes change in B; or (ii) change in B causes change in A; or (iii) A and B cause change in one another all at the same time. How can these three possibilities be discriminated?

General scientific theories often embody causal relationships, and usually specify the direction of causation. If such a theory is validated by testing it formally using real-world data, the causation is also validated.

In narrower contexts it might, alternatively, be possible to establish that a relationship is causal, by informal observation or experiment. We observe, for example, that people get sunburned more quickly in the southern hemisphere, where there is a hole in the atmospheric ozone layer, than in the northern hemisphere, where there is not. It is 'obvious', moreover, that it is the hole in the ozone layer that is accelerating sunburn, rather than the other way round. However, identifying the direction of causation is not always so straightforward. For example, when the prices of two-bedroom apartments in a particular region rise, intending buyers will generally seek to borrow more money from banks or other finance sources. However, if housing finance is more generously available, property vendors will hold out for higher prices. Is there here a dominant direction of causality?

Our naïve young man's informal experiments showed him that, when he consumes bottled drinks plus iced water, he has a hangover. (Here, we may think in terms of three interacting variables – the bottled drinks, the iced water, and the hangover – so there are several pairwise associations to consider.) He senses that there is some causation among these associations. It is obvious to him that the hangover is not causing him to drink, and he cannot (we shall assume) think of any other influence in his life that is pushing him to drink and to having hangovers. Three possibilities remain: the bottled drinks are causing the hangover; or it is the iced water; or perhaps the combination of bottled drinks and iced water. Unaware, apparently, of the critical constituent that is common to whiskey, brandy and wine, the young man decides instead (but without further experiment!) to avoid the other ingredient that is common to his experiences.

We learn from this example that appraising any set of observed associations for evidence of causality is best done in two stages. First, consider, on the basis of established (scientific) knowledge, whether each association is plausibly causal, and then investigate empirically (i.e. statistically) each plausible case and combination of cases.

Question 9.2

Many statisticians would interpret the statement 'there is a 40% chance of rain on the following day' to mean that it will rain on 40% of the days for which such a prediction is made. This is, essentially, a frequentist interpretation of the stated probability. Some statisticians might, instead, use a subjectivist interpretation – 40% is the degree-of-belief of the weather forecaster that it will rain. When pushed to explain 'degree-of-belief', they may say that this can be assessed by how willing the forecaster is to bet on his forecast turning out to be right. Further interpretations (generally offered by non-statisticians) are that it will rain on the following day for 40% of the time (that is, for almost 10 hours), or that it will rain in 40% of the area covered by the forecast. (There is more on subjectivism vs. frequentism in CHAPTER 20.)

This array of interpretations is discussed by Gerd Gigerenzer in the course of his plenary address at the Eighth International Conference on Teaching Statistics (ICOTS8), held in Slovenia in 2010. He also mentions a really inventive interpretation: from a panel of ten weather forecasters, four believe that it will rain tomorrow, while the other six disagree! A video of the address is online at [9.4]. Gigerenzer's associated ICOTS8 paper underlines the unfortunate consequences of statistical illiteracy in our society. This paper is online at [9.5].

Question 9.3

The British scientist was Francis Galton (1822–1911). The efficacy of prayer was investigated by looking at whether kings lived longer than commoners, since so many people, in effect, prayed for the king every time the words of the (British) national anthem were sung. The null hypothesis (that kings and commoners lived for the same length of time) was not rejected. Galton's paper may be read online at [9.6]. Further history is given by Brush (1974), who describes the controversy to which Galton's inquiry was a contribution.

A modern randomised double-blind trial of the efficacy of prayer is described by Byrd (1988). This study is summarised online at [9.7]. It was subsequently criticised by Tessman and Tessman (2000).

Question 9.4

Marketing strategies for commercial investments often compare money amounts at two points in time that are separated by a long interval, and highlight the magnitude of the increase over time – for persuasive impact, it is invariably an increase! Such a comparison may be entirely misleading in several ways.

a) The comparison will be misleading over the stated time period, unless the change in the value of money (i.e. the general level of price inflation) between the two points in time is taken into account. Whether by oversight or by design, this kind of adjustment of money amounts in sales promotional material is (in our experience) the exception, rather than the rule. To adjust a money amount for the effect of inflation requires an appropriate price index, and the consumer price index is not always the most appropriate. It also requires familiarity with the technique of 'deflating' money amounts using the price index, as the answer to QUESTION 7.4 illustrates.

b) The comparison may be misleading (even after adjustment for inflation) if it is implied that the same benefit will accrue from making the identical investment over an identical time period into the future. In a stochastic world, it is rarely true that the future will be substantially like the past, especially in the generally volatile world of commercial investments. Indeed, a published warning to this effect is legally required in many countries for investments marketed to the public via a formal document of offer (also called a prospectus).

c) The comparison may be misleading (even after adjustment for inflation) if it is implied that the investment being marketed now, and the investment that was successful in the past, are so alike in their nature that – just on the account of their similarity – the currently marketed investment will repeat the success of the past investment. This could be described as misleading by defective analogy. Two nearby apparently similar houses may generate quite different sales histories. The same is true for two similarly sized paintings by the same prominent artist.

Question 9.5

Direct calculation from the formula for Pearson's correlation coefficient yields $r(X,Y) = 0.22$ and $r(Y,Z) = 0.55$. It is intuitively appealing to think: 'if X is positively correlated with Y, and if Y is positively correlated with Z, then X will be positively correlated with Z'? In fact, $r(X,Z) = -0.69$. Thus, you have made a counterintuitive discovery: the direction of pairwise correlation is not (necessarily) transitive among three (or more) variables.

Allowing – and even encouraging – people to follow their intuitive, but incorrect, ideas about statistics is another way of deliberately producing misleading statistics. QUESTION 10.2 is a further example of this.

An (advanced) algebraic proof of the nontransitivity of pairwise correlations in general is given in Langford, Schwertman and Owens (2001). The data in this question are adapted from that paper. Students' misapprehension of this counterintuitive result is explored in Castro Sotos *et al.* (2009), online at [9.8].

References

Print

Brush, S.G. (1974). The prayer test. *American Scientist* **62**, 561–563.

Byrd, R.C. (1988). Positive therapeutic effects of intercessory prayer in a coronary care unit population, *Southern Medical Journal* **81**, 826–829.

Langford, E., Schwertman, N. and Owens, M. (2001). Is the property of being positively correlated transitive? *The American Statistician* **55**, 322–325.

Tessman, I. and Tessman, J. (2000). Efficacy of prayer. *Skeptical Inquirer* **24**(1), 31–36.

Online

[9.4] http://videolectures.net/icots2010_gigerenezer_hdap/(*Note*: this link misspells the name Gigerenzer)

[9.5] Gigerenzer, G. (2010). Helping doctors and patients make sense of health statistics: towards an evidence-based society. In Reading, C. (ed.). *Proceedings of ICOTS8*. International Statistical Institute. At http://www.stat.auckland.ac.nz/~iase/publications/icots8/ICOTS8_PL3_GIGERENZER.pdf

[9.6] Galton, F. (1872). Statistical inquiries into the efficacy of prayer. *Fortnightly Review* (new series) **68**, 125–135. At http://www.abelard.org/galton/galton.htm

[9.7] http://www.medicine.ox.ac.uk/bandolier/band46/b46-6.html

[9.8] Castro Sotos, A., Vanhoof, S., van den Noortgate, W. and Onghena, P. (2009). The transitivity misconception of Pearson's correlation coefficient. *Statistics Education Research Journal* **8**, 33–55. At http://www.stat.auckland.ac.nz/~iase/serj/SERJ8(2)_Sotos.pdf

Answers – Chapter 10

Question 10.1

a) P(A) = 4/52 = 1/13. P(B) = 13/52 = 1/4. P(AB) = P(ace and spade) = 1/52. Then P(AB) = P(A).P(B), and so the events are statistically independent. This is true by definition. The fact that there is a logical connection (or dependence) between drawing an ace and drawing a spade, in the sense that there can be an outcome that is both an ace and a spade, is irrelevant to the statistical independence of events A and B.

b) Yes, the converse is true also. For example, call A the event that a student has gained a total greater than 490 out of 500 over all sections of a national entrance exam for undergraduate study to a country's universities. Call B the event that a student is doing undergraduate study at the highly-regarded University of X. Then, in principle, P(A|B) > P(A). However, the statements defining A and B are logically unconnected, in the sense that neither implies the other.

Question 10.2

When three coins are tossed once, the probability that each shows a head is $(0.5)^3 = 0.125$. The probability that each shows a tail is the same. Because the event 'three heads' and the event 'three tails' are mutually exclusive, the probability that the coins all show the same face is $0.125 + 0.125 = 0.25$.

   Where is the logical flaw in the alternative (incorrect) line of reasoning that leads to the answer 0.5? It can be detected most directly by enumerating the eight possible cases:

   (H, H, H), (H, H, T), (H, T, H), (H, T, T), (T, H, H), (T, H, T), (T, T, H), (T, T, T).

   Consider now the statements, 'when three coins are tossed, two of them must show the same face. The third coin will show a head or a tail, in either case with probability 0.5.' The first statement is clearly true. However, the second is false. To see this, focus on the cases with two heads. Four of the above-listed cases include two heads. In three of these cases, the third face is a tail, but in only one case is it a head. So it is, in fact, *three times as likely* that the third coin will be a tail as a head.

   The crux of the paradox is that the argument that produces the incorrect answer subtly induces the mind to limit attention to the outcomes (H, H, H), (H, H, T), (T, T, H) and (T, T, T) alone.

   This paradox is presented by Northrop (1961), page 165 (online at [10.1]), who traces its origin to a paper by Francis Galton (1894).

Question 10.3

a) Assuming first that a year has 365 days, each equally likely to be a birthday (thus ignoring the observed fact that, in many countries, there are more births in some months than in others), the probability of *no* match in the birthdays of N randomly selected people is:

$$U_N = \left(1 - \frac{1}{365}\right)\left(1 - \frac{2}{365}\right)\cdots\left(1 - \frac{N-1}{365}\right) = \frac{364!}{N!365^{N-1}} = \frac{365!}{N!365^N} = {}^{365}P_N / 365^N$$

   Then the chance that there will be at least one matching birthday among these N people is $V_N = 1 - U_N$. Taking N as 22 and 23 gives values for V_N of 0.476 and 0.507, respectively.

   For a year of 366 days, we replace 365 by 366 in the formula for V_N (while admitting that the assumption that each day of a leap year is equally likely to be a birthday is implausible). Then, for N equal to 22 and 23, V_N is 0.475 and 0.506, respectively.

   Hence, the smallest number of people for at least an even chance of a match in birthdays is 23, whether or not the year is a leap year.

There is an extensive literature on this counterintuitive result, some of it exploring variants of the original problem. Falk (2014) is of interest, because she includes a discussion of the question why intuitive answers to the problem deviate so much from the correct answer.

b) FIGURE 26.1 shows the birth dates of British Prime Ministers, arranged in reverse chronological order of their first year as PM, beginning with David Cameron, who was PM at the time of writing. The dates can be found online at [10.2]. The table reveals that we need to consider 28 PMs to find a match in birth dates. Both John Major and Edward Smith-Stanley have the birth date 29 March.

David Cameron	9 Oct	1966
Gordon Brown	20 Feb	1951
Tony Blair	6 May	1953
John Major	**29 Mar**	1943
Margaret Thatcher	13 Oct	1925
James Callaghan	27 Mar	1912
Edward Heath	9 Jul	1916
Harold Wilson	11 Mar	1916
Alec Douglas-Home	2 Jul	1903
Harold Macmillan	10 Feb	1894
Anthony Eden	12 Jun	1897
Clement Attlee	3 Jan	1883
Winston Churchill	30 Nov	1874
Neville Chamberlain	18 Mar	1869
Ramsay MacDonald	12 Oct	1866
Stanley Baldwin	3 Aug	1867
Andrew Bonar Law	16 Sep	1858
David Lloyd George	17 Jan	1863
Herbert Asquith	12 Sep	1852
Henry Campbell-Bannerman	7 Sep	1836
Arthur Balfour	25 Jul	1848
Archibald Primrose, Lord Rosebery	7 May	1847
Robert Cecil, Marquis of Salisbury	3 Feb	1830
William Gladstone	29 Dec	1809
Benjamin Disraeli	21 Dec	1804
Henry Temple, Lord Palmerston	20 Oct	1784
George Hamilton-Gordon, 4thEarl of Aberdeen	28 Jan	1784
Edward Smith-Stanley, 14th Earl of Derby	**29 Mar**	1799

Figure 26.1 Birthdates of British Prime Ministers.

c) The fundamental distinction between the 'birthday problem' and the 'prime ministers problem' is that in the former we are looking at a *fixed number* of people and asking for the probability that there is at least one match, while in the latter we are looking at a situation where there is an *indeterminate number* of people, whom we are examining one by one until a particular condition is satisfied (namely, that the first match occurs).

Feller (1968) discusses the birthday problem (page 33) and (the general setting of) the prime ministers problem (page 47); then he shows (page 49) that these two problems, while theoretically distinct, have some numerically identical characteristics. Thus, the probability that a fixed number (call it k) of people all have distinct birthdays in the birthday problem is the same as the probability that one will not find a match of birthdays after examining k people's birthdays in the prime ministers problem.

A theoretical analysis of the prime ministers problem follows in the answer to QUESTION 10.4.

Question 10.4

a) Let N be the number of birthdays to examine to find the first match (possible values from 2 to 365). Using a standard approach for calculating the expected value of a random variable, we write N as a sum of 'indicator' variables: $N = 1 + N_1 + N_2 + N_3 + ... + N_{365}$ where $N_n = 1$ if the first n birthdays are distinct, $= 0$ otherwise. Now,

$$E(N_n) = \Pr(N_n = 1) = {}^{365}P_n / 365^n = {}^{365}C_n n! / 365^n, \text{ and so}$$

$$E(N) = \sum_{n=0}^{365} \left[{}^{365}C_n n! / 365^n \right].$$

This calculation is best carried out by a computer! It can be done directly in a program such as *Mathematica*, where the appropriate expression is:

$$N\text{Sum}\left[\text{Binomial}[365, n] \times n! / 365^n, \{n, 0, 365\} \right]$$

giving the result 24.6166.

In the (free) online *WolframAlpha* [10.3], a little experimentation shows that the expression Sum(Combin(365,n)*n!/365^n) $n=0$ to 365 also results in 24.6166.

In *Excel*, there is a problem with direct evaluation, as $n!/365^n$ gets very large quickly, and by $n = 121$ it is being reported as [#NUM!] rather than a numerical value. An alternative approach is to express each term of the sum as a multiple of the previous term, obtaining the recursion relationship $T_{n+1} = T_n (365 - n)/365$. Since $T_0 = 1$, it follows that $T_1 = 1$, $T_2 = 1 \times (364/365) = 0.997$, etc. With this approach, *Excel* can be used to find the sum, given as 24.6166 to four decimal places and, indeed, this result is obtained if the sum is taken only to $n = 100$ rather than all the way to $n = 365$.

b) In the case of Australia (which has had 29 Prime Ministers since Federation in 1901, up to the year of writing, 2016), going back in order from the current PM, a match of birthdays is found only after all 29 birthdays have been examined – see the data online at [10.4]. The 24th PM, Paul Keating, has the same birthday (18 January) as the very first PM, Edmund Barton.

For US Presidents, going back in order from the current President (in early 2016), Barack Obama, a match is found after 33 birthdays have been examined. Warren Harding (in office 1921–23) and James Polk (in office 1845–49) both have their birthday on 2 November.

Question 10.5

a) The full title of the book is *Pillow Problems Thought Out During Sleepless Nights*. The author of this work is shown on the title page as Lewis Carroll. This was the pen name of the English mathematician and writer Charles Lutwidge Dodgson (1832–1898). Dodgson described the creation of his pen name as follows: his first names Charles Lutwidge, translated into Latin, become Carolus Lodovicus. Transposing these Latin names, and retranslating freely into English, produces Lewis Carroll. *Pillow Problems* has been republished as Carroll (1958), in a collection of his writings with the title *Mathematical Recreations of Lewis Carroll*.

A short biography of Carroll, and a summary of his mathematical contributions, may be found online at [10.5]. Nowadays, he is known worldwide not for his mathematical works, but for his vivid books for children, including *Alice's Adventures in Wonderland*, *Through the Looking-Glass* and *The Hunting of the Snark*.

The 13 probability puzzles among the Pillow Problems are reviewed and dissected in Eugene Seneta's enjoyable essay, Seneta (1984). A broader overview is in Seneta (1993), online at [10.6].

b) This is the fifth of the 72 Pillow Problems. The hint given in the question was not supplied by Lewis Carroll but, without it (as later writers have pointed out), the problem is incompletely formulated. A neat solution is reproduced from a reader's letter by Martin Gardner on page 189 of his book (Gardner, 1981). The following is quoted from this source: 'Let B and W(1) stand for the black or white counter that may be in the bag at the start and W(2) for the added white counter. After removing a white counter there are three equally likely states – W(1) in bag and W(2) outside; W(2) in bag and W(1) outside; B in bag and W(2) outside. In two of these states a white counter remains in the bag, and so the chance of drawing a white counter the second time is 2/3.'

A naïve approach to the problem, which observes that the state of the bag at the end of the experiment is the same as it was at the beginning and, therefore, leads to the incorrect answer 1/2, fails to take into account the conditioning that is essential to arriving at the correct answer.

A note on incompletely formulated probability problems: Dodgson is not the only mathematician who sought to invent challenging probability puzzles, only to have it made clear later that some of the headaches these problems caused others were the result of incomplete or ambiguous wording in their initial formulation which the inventor overlooked. It is worth bearing this in mind when a probability problem next gives you a headache!

Three such problems, including another one of Dodgson's *Pillow Problems* ('the obtuse problem'), are analysed by Martin Gardner in chapter 19 ('Probability and Ambiguity') of Gardner (1966). Ruma Falk and her colleagues have devoted a lot of care to revealing the ambiguity embedded in many familiar probability problems. See, for example, Bar-Hillel and Falk (1982), Falk and Samuel-Cahn (2001) and Falk and Kendig (2011). There is a further incompletely formulated probability problem in QUESTION 20.3.

References

Print

Bar-Hillel, M. and Falk, R. (1982). Some teasers concerning conditional probabilities. *Cognition* **11**, 109–122.

Carroll, L. (1958). *Mathematical Recreations of Lewis Carroll – Pillow Problems and a Tangled Tale*. Dover.

Falk, R. and Samuel-Cahn, E. (2001). Lewis Carroll's obtuse problem. *Teaching Statistics* **23**, 72–75.

Falk, R. and Kendig, K. (2011). A tale of two probabilities. *Teaching Statistics* **35**, 49–52.

Falk, R. (2014). A closer look at the notorious birthday coincidences. *Teaching Statistics* **36**, 41–46.

Feller, W. (1968). *An Introduction to Probability Theory and Its Applications*, Volume 1, 3rd edition. Wiley.

Galton, F. (1894). A plausible paradox in chances. *Nature* **49**(1268), 365–366.

Gardner, M. (1966). *More Mathematical Puzzles and Diversions*. Penguin.

Gardner, M. (1981). *Mathematical Circus*. Allen Lane.

Seneta, E. (1984). Lewis Carroll as a probabilist and mathematician. *Mathematical Scientist* **9**, 79–94.

Online

[10.1] Northrop, E.P. (1961). *Riddles in Mathematics*. Penguin. At https:// archive.org/details/RiddlesInMathematics

[10.2] http://en.wikipedia.org/wiki/ List_of_Prime_Ministers_of_the_United_Kingdom_by_longevity

[10.3] https://www.wolframalpha.com/

[10.4] http://en.wikipedia.org/wiki/ List_of_Australian_Prime_Ministers_by_age

[10.5] http://www-groups.dcs.st-and.ac.uk/~history/Biographies/Dodgson. html

[10.6] Seneta, E. (1993). Lewis Carroll's 'Pillow Problems': on the 1993 centenary. *Statistical Science* **8**, 180 -186. At https://projecteuclid.org/ euclid.ss/1177011011

Answers – Chapter 11

Question 11.1

Sopra le scoperte dei dadi ('On the outcomes of [rolling] dice') is a short essay written by Galileo Galilei (1564–1642). It is of uncertain date, but generally thought to be sometime between 1613 and 1630. Galileo wrote (in the English translation cited below): 'Using three dice, 9 and 10 points can each be obtained in six ways. How is this compatible with the experience that, based on long observation, dice players consider 10 to be more advantageous than 9?' His essay resolved this long-standing gambling problem.

Galileo used the notation (123) to mean a 1 spot on one die, a 2 spot on another die and a 3 spot on a third die. Then, if order is *not* taken into account, a total of 9 can indeed be obtained in six ways: (126), (135), (144), (234), (225), (333). A total of 10 can also be obtained in six ways: (145), (136), (226), (235), (244), (334). However, Galileo reasoned insightfully that *order must be taken into account* for correctly determining relative frequencies of occurrence. Thus, 10 is actually obtainable in 27 ways, while 9 is obtainable in only 25 ways. Since the total number of outcomes (taking account of order) when three dice are rolled is $6^3 = 216$, the probability of rolling 10 is $27/216 = 0.125$, which is marginally greater than the probability of rolling 9, which is $25/216 = 0.116$. In this brief pioneering excursion into the quantification of chance, Galileo anticipated by at least 25 years a general principle formulated in 1654 by Fermat, in correspondence with Pascal, for the correct calculation of probabilities of compound events.

An English translation of Galileo's essay can be found in Appendix 2 (pages 192–195) of David (1962). Galileo's solution is discussed on pages 64–66 of the same book, and also on pages 239–240 of Freedman, Pisani and Purves (2007).

Question 11.2

The example of a multiplicative congruential generator (MCG) we gave in the Overview was $X_{n+1} = 11X_n \pmod{13}$, with $X_0 = 9$. This yields the following sequence of pseudorandom values, beginning at X_1: 8, 10, 6, 1, 11, 4, 5, 3, 7, 12, 2, 9, 8, 10 ... Observe that this sequence is periodic, with period length 12, and that all integers in the range 1–12 are present in a 'pseudorandom' order. All MCGs are periodic, though the period length is not always maximal. (Because there is no 0 in the above sequence, this example does not have the maximum possible period length of an MCG with modulo 13 – namely, 13.) Clearly, an MCG with a short period will not be very useful as a pseudorandom number generator in practice. We must choose an MCG with values for the multiplier, modulo and seed that jointly guarantee a period longer than the total number of 'random numbers' we need, as well as producing a sequence that passes standard tests of 'patternlessness'. There are mathematical theorems to guide us to suitable choices.

However, there is one systematic pattern in the sequence of values generated by any MCG (within its period) that is inescapable. This pattern is revealed if we plot the overlapping successive pairs of generated values on a graph. A plot of the 12 points with (X,Y) coordinates (8,10), (10,6), (6,1), (1,11) ... (9,8) shows that they all lie on one of two parallel lines in the (X,Y) plane. Every MCG produces values that, when plotted in this fashion, lie *without exception* along a relatively small number of parallel straight lines in the (X,Y) plane, leaving other areas devoid of points. If the generated values were truly random, that would not happen: the points would scatter evenly over the *entire* (X,Y) plane. An alternative way of stating this is that successive pairs of MCG-generated pseudorandom numbers are linearly correlated and, thus, are not statistically independent. It is possible to design an MCG so that this correlation, over the full period of the generator, is quite small – but it will never be zero.

Question 11.3

The author is George Spencer Brown, and the passage quoted is from page 57 of his book *Probability and Scientific Inference* (1957). Several chapters of this very readable, but quite unsettling, book are devoted to penetrating discussions of the meaning of randomness and the validity of defining the probability of an event, in terms of the frequency of occurrence of that event in a sequence of random trials.

Spencer Brown points out first an ambiguity in conventional thinking about random sequences. We say that a sequence of digits is random (in the abstract) if it is 'patternless', and if any particular digit in the sequence is not predictable from knowing the digits that have been generated before. We also say that a sequence of digits is random (in operational terms) if it has satisfactorily passed a predefined set of statistical (and, perhaps, other) tests of randomness. Clearly, neither of these definitions implies the other.

Focusing next on the operational notion of a random sequence, in the context of (0,1) random digit generation, Spencer Brown shows that it is liable to produce the self-contradiction highlighted in the passage quoted in this question. This self-contradiction, in turn, points to a paradox. Whether a sequence is called random is not only a property of the entire sequence that the generator might generate (call it 'global randomness'); it depends also on how much of the sequence we have actually inspected (call it 'local randomness').

To resolve this paradox, we could imagine (suggests Spencer Brown) interposing a metering device that regulates the maximum number of noughts in a row that a random generator is allowed to produce for a given sequence length, *and permits that number to increase progressively* as the generated sequence lengthens. Suppose it is agreed – as one of the criteria for randomness in a sequence – that, at most, six noughts in a row are permitted by the meter if the sequence is less than 30 digits long. Then if, say, the 15th to the 20th digits generated are noughts, we know for certain that the 21st digit must be a one (because the meter is operating). In that case, the 21st digit cannot be said to be randomly generated. Thus, this attempted resolution of the paradox has failed.

After further unsuccessful approaches to resolving the paradox, Spencer Brown concludes that no resolution is possible: *it is actually not a paradox at all*. In other words, the general concept of a random sequence is simply not operationally definable in any consistent way. (This has gloomy consequences, of course, for the practical viability of the relative frequency definition of probability.)

Some years later, the probabilist Mark Kac (1914–1984) came to a similar conclusion, but offered this (paradoxical) reassurance: 'The discussion of randomness belongs to the foundations of statistical methodology and its applicability to empirical sciences. Fortunately, the upper reaches of science are as insensitive to such basic questions as they are to all sorts of other philosophical concerns. Therefore, whatever your views and beliefs on randomness … no great harm will come to you. If the discipline you practice is sufficiently robust, it contains enough checks and balances to keep you from committing errors that might come from the mistaken belief that you *really* know what "random" is.' (Kac, 1983).

Question 11.4

Given three random events, X, Y and Z, it is intuitively appealing (but false!) to believe that, if X is more likely than Y, and Y is more likely than Z, then it must be the case that X is more likely than Z – in other words, that probability orderings are transitive. These four dice are interesting because they can be used to illustrate the nontransitivity of probability orderings.

When these dice are rolled, A is likely to yield a higher face number than B, B than C, C than D, *but D is likely to yield a higher number than A!* In each case, the long-run probability is 2/3. To see this in the first case, for example, write the face numbers on dice A and B across the top and down the side, respectively, of a 6 × 6 array. Fill the array with + and − symbols, according as (A−B) is positive or negative. 24 cells will be found to contain a +, showing that, in the long run, the A face number will exceed the B face number with probability 2/3.

Another way of putting this is as follows: if you and your friend are playing a game in which you each roll one of the dice and the higher number wins then, after your friend has selected his or her die, you can always select another die that will give you the greater (long-run) chance of winning.

The original idea of nontransitive dice is due to the US statistician Bradley Efron, although this particular set was designed by the US physicist Shirley Leon Quimby (1893–1986). Efron's nontransitive dice are shown by Martin Gardner in one of his Mathematical Games columns in *Scientific American* (Gardner, 1970). A web search will turn up many other examples of sets of nontransitive dice.

Transitivity in a different context is investigated in QUESTION 9.5.

Question 11.5

Feller (1968) discusses the coin-tossing game on pages 78–88 of chapter 3 ('Fluctuations in coin tossing and random walks') in his book. He writes (page 78): 'According to widespread beliefs a so-called law of averages should ensure that in a long coin-tossing game each player will be on the winning side for about half the time, and that the lead will pass not infrequently from one player to the other.' However, 'the amazing fact' is that these propositions – both so intuitively plausible – are incorrect.

Feller proceeds to explain why by analysing an equivalent situation – a symmetric random walk, starting at the origin, in which a person takes steps independently, with equal chance to the right or the left. We can say that a 'head' corresponds to a step to the right, and a 'tail' to a step to the left. A is ahead if the current position is to the right of the origin, and B is ahead if the current position is to the left of the origin. Feller demonstrates (page 82) the surprising result that the most likely value for the proportion of time spent on the right side of the origin is 0 or 1, and the least likely result is 0.5. It follows that it is naïve to suppose that A will be ahead around 500 times in any particular game of 1000 tosses.

The U-shaped distribution of the proportion of time that person A is ahead is referred to as the (discrete) arcsine distribution. (Incidentally, it is a further statistical example of a U-shaped distribution, as mentioned in the answer to QUESTION 5.5.)

Conventionally, if the probability of an event or of one that is more extreme is below 0.05, we conclude that it is unlikely to have occurred by chance. Is A's experience of being ahead for just 50 tosses in 1000 such an extreme event (suggesting, for example, that the coin may be biased)? Feller calculates (page 81) that the chance of A being ahead for 200 tosses or fewer is 0.295, while for 100 tosses or fewer, it is 0.205, and for 50 tosses or fewer it is 0.144. Only when A is ahead for six tosses or fewer does the probability fall to 0.049. Thus, even though A may be surprised at being ahead for only 50 tosses in 1000, this still suggests that she is playing in a fair game.

As to the intuitive proposition that 'the lead will pass not infrequently from one player to the other', Feller's analysis shows (page 81) that 'contrary to popular notions, it is quite likely that, in a long coin-tossing game, one of the players remains practically the whole time on the winning side, the other on the losing side.'

Feller sums up (page 72): 'if even the simple coin-tossing game leads to paradoxical results that contradict our intuition, the latter [i.e. our intuition] cannot serve as a reliable guide in more complicated situations.' As we pointed out at the beginning of CHAPTER 10, even statisticians find probability theory difficult!

An interesting application of these results forms the basis of a paper by Denrell (2004). The author considers a situation where Firm 1 consistently outperforms Firm 2 in profitability over time. This difference in performance is usually attributed to characteristics of the firms, with the conclusion that Firm 1 is better than Firm 2. However, it may be that there is no real difference between the firms, and that such a consistent difference is an instance of the chance phenomenon of long leads in random walks. In the world of business (as elsewhere), spurious theories may be put forward to account for essentially random phenomena.

References

Print

David, F.N. (1962). *Games, Gods and Gambling – A History of Probability and Statistical Ideas*. Griffin.

Denrell, J. (2004). Random walks and sustained competitive advantage. *Management Science* **50**(7), 922–934.

Feller, W. (1968). *An Introduction to Probability Theory and Its Applications*, Volume 1, 3rd edition. Wiley.

Freedman, D., Pisani R. and Purves R. (2007). *Statistics*, 4th edition. Norton.

Gardner, M. (1970). The paradox of the nontransitive dice. *Scientific American* **223**(6), 110–114. Reprinted with updates as chapter 5 in Gardner, M. (1983). *Wheels, Life, and Other Mathematical Amusements*. Freeman.

Kac, M. (1983). What is random? *American Scientist* **71**, 405–406.

Spencer Brown, G. (1957). *Probability and Scientific Inference*. Longmans Green.

Answers – Chapter 12

Question 12.1

In the casino game Blackjack (also known as Vingt-et-un or Twenty-one), the gambler aims to score a higher total with his or her cards than the dealer, but no more than 21. Face cards (kings, queens, jacks) are counted as 10, an ace can be counted as 1 or 11, and other cards are counted at face value. A successful gambler is paid the amount of his or her original stake, plus the return of the stake. However, if the gambler and the dealer have the same point total, this is called a 'push', and the gambler neither wins nor loses money on that bet. The gambler could then retrieve the stake and quit, or could use it to bet on the next round.

Question 12.2

Despite the symmetry in the description, the game is not fair. This may come as a surprise. Even more surprising, then, might be the discovery that (despite all the card shuffling) this is not actually a gambling game at all – the end result is always the same! Why? Each win will multiply your current pot by 1.5, and each loss will multiply it by 0.5. Five wins and five losses, *regardless of order*, will multiply your original pot by $(1.5)^5(0.5)^5 = 0.2373$, leaving you with $23.73 – a loss of more than 75%.

If you were to bet the same *amount*, rather than the same *proportion of your pot*, at each stage – say, $10 – the game would be fair (but the outcome is still deterministic). Over the course of the ten plays, your initial pot would increase by 5 × $10 and decrease by 5 × $10.

The game described in the question would become a gamble if the 10 cards were shuffled again after each card is turned over (equivalent to 'sampling with replacement'). In this version, the game would be fair, since your expected pot after each card is turned over would be $100 × (0.5 × 1.5 + 0.5 × 0.5) = 100$.

A deeper look at this gamble highlights a fundamental truth about many commercially-offered gambles. That truth is revealed when we ask: how does the complete probability distribution of your winnings look if you play this game many times?

Perhaps surprisingly, it is skewed. You have a high chance of losing some of your money, though you cannot lose more than your initial $100. On the other hand, you have a low chance of making a lot of money. For instance, if you play ten times, you will have a probability of 0.83 of losing money (if a black card is turned over 0–6 times), but you would win money if a black card were turned over seven or more times – as much as $5000, if a black card were turned over ten times. If you play 50 times, the results are even more extreme. You have a 0.97 probability of losing money (if a black card is turned over up to 31 times), and only a 0.03 probability of winning money. Indeed, you would make more than $1 million if a black card were turned over 43 times (or more). These probability calculations are based on the binomial distribution.

This asymmetry – a large chance of losing a small amount and a very small chance of winning a large amount – is what is so characteristic of the gambling 'industry'.

This question is also discussed on pages 72–73 of chapter 6 ('Random Walks and Gambling') in Gardner (1981), and by Nalebuff (1989).

Question 12.3

The author referred to is Sir William Petty (1623–1687), an economist and a founding member of The Royal Society, who pioneered the field of 'political arithmetic' (or economic statistics, as we call it today). His name is prominent in the history of statistical ideas (see FIGURE 22.1).

Here is Petty's blunt assessment of lotteries (online at [12.4]): 'Now in the way of lottery men do also tax themselves in the general, though out of hopes of advantage in particular. A lottery therefore is properly a tax upon unfortunate self-conceited fools ... Now because the world abounds with this kinde of fools, it is not fit that every man that will, may cheat every man that would be cheated; but it is rather ordained, that the sovereign should have the guardianship of these fools, or that some favourite should beg the sovereign's right of taking advantage of such men's folly, even as in the case of lunaticks and idiots.'

Question 12.4

The first of Adam Smith's two assertions is, in practice, always correct: if you buy all the tickets in a lottery you will win all the prizes but they will, in total, be of lower value than the amount you have spent – otherwise, the people running the lottery would make no money. The second assertion appears incorrect: if you buy an extra ticket it would, in fact, increase your chance of winning a prize, *until you had bought so many tickets that you had paid out more than the total of the available prizes.* Perhaps Smith was thinking of a particular context where ticket prices are unequal, so that the tipping point specified in the italicised words of our previous sentence might be reached with the acquisition of only relatively few tickets.

Alternatively, Smith may have been thinking intuitively, for there is a frame of reasoning – unrecognised in Smith's day – in which his statement, slightly adapted, makes sense. Our adaptation is to switch attention from a focus on chance (i.e. the *probability* of a return from buying tickets) to a focus on the money amount that might be won or lost as a result of the play of chance (i.e. the *expected return* from buying tickets), and then to reason in terms of expected return.

Suppose 1000 tickets are sold at £1 each in a lottery where there is only one prize – £500. Then the expected return if you buy N tickets is defined by 'N chances in 1000 of winning £500 less £N, the cost of the tickets'. In symbols we write: expected return (in £) $= 500(N/1000) - N = -(N/2)$. In a commercial lottery, the expected return will *always* be negative and, the more tickets you buy, the greater will be your expected loss in money terms. In this example, the expected loss will rise to £500 when you buy all the tickets.

Question 12.5

As noted in the Overview, the result of repeated straight-up bets at roulette can be modelled using the binomial distribution. Let n be the number of bets made, and $p = 1/37$ the success probability. Then, the number of wins, X, will have a binomial (n, p) distribution. Since each bet costs $1 and each successful bet returns $36 (including the original $1 bet), your total profit from n bets will be $Y = 36X − n$, and you will be ahead if this is positive. Now let us consider the probability of being ahead as a function of n, the number of plays.

Important preliminary: *it is because the number of plays is a discrete variable that the paradox highlighted in the question arises.*

To be ahead at any time during the first 35 plays, you need to win only once. Each extra play will give you an added chance of winning, so your probability of being ahead will increase with n. During the plays from 36 to 71, however, you will need two wins to be ahead. Thus, when you make the 36th play, your probability of being ahead will go down. For plays from 37 to 71, the probability will increase, but it will go down again when you make the 72nd play, since you will now need three wins to be ahead. As a function of the number of plays, the chance of being ahead increases for every play, except for those that are a multiple of 36. At these points, your chance of being ahead decreases.

Using the binomial distribution, we have calculated the numerical probabilities of being ahead: (a) at an exact multiple of 36 plays; and (b) at one less than the next multiple of 36. These probabilities are graphed in FIGURE 26.2 for selected multiples of 36 ('circles'), and for 35 plays later, just before the next multiple of 36 ('triangles'). As the number of multiples of 36 increases, you can see that each of these sets of probabilities forms a decreasing sequence, reflecting the basic gambling truth that 'the longer you play, the less likely you are to be ahead'.

FIGURE 26.3, which plots the binomial probability of being ahead for all values of n between 500 and 600, lets you see in close-up how this probability behaves. The function displays a sawtooth pattern, increasing for every value of n except for the ones that correspond to the next multiple of 36, when the probability drops. This illustrates clearly that most of the time, playing one extra game will increase your chances of coming out ahead.

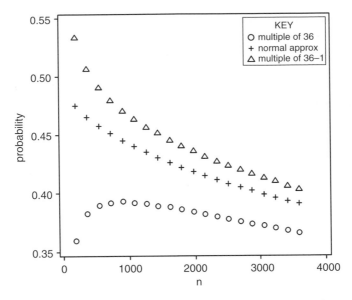

Figure 26.2 Exact and approximate probability of being ahead at roulette for selected plays.

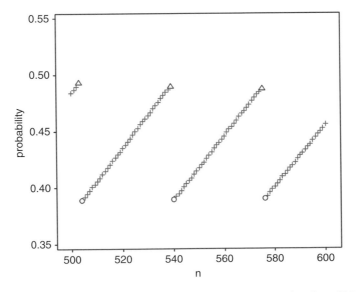

Figure 26.3 Probability of being ahead at roulette for consecutive plays from 500 to 600.

Note: the normal approximation to the binomial probability of being ahead treats the number of plays as a continuous variable. Then the (approximate) probability of being ahead has no sawtooth pattern; rather, it declines monotonically. This can be seen graphically in FIGURE 26.2, where the normal approximation to the binomial probability of being ahead is plotted with crosses.

References

Print

Gardner, M. (1981). *Mathematical Circus*. Allen Lane.
Nalebuff, B. (1989). Puzzles. *Journal of Economic Perspectives* **3**(3), 166 and 169.

Online

[12.4] Petty, W. (1662), *A Treatise of Taxes and Contributions*, chapter 8. At http://quod.lib.umich.edu/e/eebo/A54625.0001.001/1:11?rgn=div1;vid=98850;view=fulltext

Answers – Chapter 13

Question 13.1

The coin might land on its edge – very possible if it were tossed in a muddy field – or it may not land at all – conceivable if the glint of its spinning attracts a hungry passing bird! To each of these outcomes, textbooks generally assign zero probability. From this, two things become clear: firstly, that textbook models of statistical experiments (like all models of real-world processes) are always, in some way, abstractions of reality; and, secondly, that an assigned probability of zero does not automatically mean that an outcome is impossible. In other words, zero probability is a necessary, but not sufficient, condition for impossibility.

Question 13.2

Two obvious shape characteristics of the normal distribution are its symmetry and the 'thinness' of its tails – only 1.2% of the area under any normal curve lies outside the range, mean ± 2.5 standard deviations (see also QUESTION 14.1(c)). The normal distribution will be a practically suitable model for real-world data that have a histogram that approximates these shape characteristics.

The brilliant mathematician Carl Friedrich Gauss (1777–1855) was one of the first to investigate the approximate correspondence of the normal distribution to real-life phenomena. Gauss surmised that any physical or biological real-life variable is likely to be symmetrically distributed, with quite 'thin' tails, if its values are determined by a large number of independent random causes, each individual cause having only a small role in the process. This quite aptly describes the processes that determine the variation of individual adult heights around the population mean height for adults *of a single gender*. Thus, the distribution of a population of male or female adult heights can be expected to resemble a normal distribution, provided that the population: (a) is not limited in any way that is implicitly height-related (for example, only professional basketball players); and (b) is large, so that there is a good chance that some very short and some very tall individuals will be represented.

What if the population comprises both genders? In most societies, women have a smaller mean height than men. Given that the normal distribution is a suitable model for the heights of each gender individually, a population composed of both genders would be modelled by a *mixture* of two normals with different means (and, perhaps also, different standard deviations). Would such a mixture of normals appear bimodal, or might it turn out unimodal? This puzzle is investigated in Schilling *et al.* (2002).

Question 13.3

The probability model is the Poisson distribution, named after the French mathematician Siméon-Denis Poisson (1781–1840). In the years 1835–37, Poisson investigated the effect of the rule for jury verdicts (either unanimity or some particular required majority), and of jury size on the probability of a correct verdict. His approach is outlined on pages 186–194 in Stigler (1986). This was one of the earliest appearances of the probability model that bears his name.

Between September 1944 and March 1945, London was bombarded by more than a thousand V2 rockets, launched by German forces from continental Europe. These rockets travelled faster than the speed of sound and so struck without audible warning, causing civilian deaths on a large scale. It was immediately a matter of importance to the British authorities, seeking to protect civilians, to know whether these rockets were guided missiles or whether they struck the ground at random. It proved impossible for quite some time to find an unexploded rocket and inspect it directly.

A different approach was needed, and a probability model provided that approach.

After several months of the rocket barrage, when 537 rockets had already fallen on South London, a statistical analysis involving the Poisson distribution was undertaken to discover whether the rockets had a guidance mechanism. A 24 × 24 grid of lines, 500 metres apart, was superimposed on a map of South London, and the number of V2 rockets striking each of the 576 resulting map squares was counted. The good fit of the Poisson model to the observed frequency distribution of rocket strikes per map square enabled the authorities to conclude that the rockets were falling at random – or, in other words, that they could not be aimed precisely. After the war ended, the data and calculations from this study were made public by Clarke (1946), online at [13.2].

Question 13.4

a) The exponential distribution – which, confusingly, some writers call the negative exponential distribution – has probability density function $f(y) = \lambda e^{-\lambda y}$ (with $y \geq 0$) and distribution function $F(y) = \Pr(Y \leq y) = 1 - e^{-\lambda y}$. This distribution is commonly used as a probability model for the length of 'service time' in the statistical study of queues. The exponential distribution has the (mathematical) property of 'memorylessness'. In the queuing context, this means that, having already spent s minutes being served, a person has the same probability of a *further* t minutes of service as a person who has just begun to be served. In symbols, if Y represents the time being served, then $\Pr(Y > s + t \mid Y > s) = \Pr(Y > t)$ for $s > 0$ and $t > 0$. The proof is direct:

$$\Pr(Y > s + t \mid Y > s) = \Pr(Y > s + t)/\Pr(Y > s) = e^{-\lambda(s+t)}/e^{-\lambda s} = e^{-\lambda t} = \Pr(Y > t).$$

Service times are not 'memoryless' in *all* real-world queues (for example, queues at traffic lights), but this property turns out to be realistic enough to make it a useful assumption in a broad class of practical situations. The exponential distribution is the only continuous probability distribution with the memoryless property – hence its importance in this applied context. For more detail on the memoryless property, see Vaughan (2008), online at [13.1].

b) If we assume that 'memoryless' service time is a realistic assumption in this queuing problem, we can find the solution immediately. When the first of the currently-served customers (say A) finishes being served, and C starts to be served, the other customer, B, has the same distribution for the time that she continues being served as does C. Hence, the probability that B finishes before C – and, thus, that C is the last to leave – is 0.5.

Question 13.5

If the proposition (which comes from a student's mistake at an examination) were correct, then it would follow that the probability of three arrivals in three hours is three times the probability of one arrival in one hour, that the probability of four arrivals in four hours is four times the probability of one arrival in one hour, and so on. Eventually, irrespective of the value of the Poisson mean arrival rate per hour, λ, the 'probability' would exceed one, which is obviously nonsensical. Hence, it seems reasonable that the probability of two arrivals in two hours is less than twice the probability of one arrival in one hour.

Let's take this analysis further. Why might it be suggested that the probability of two arrivals in two hours is twice the probability of one arrival in one hour? Adding probabilities is correct only in the context of finding the probability of an event which is the union of two mutually exclusive events. What might these mutually exclusive events be? Perhaps the student was thinking that one arrival in the first hour and one arrival in the second hour are mutually exclusive – but, clearly, they are not. In fact, under the assumption of Poisson arrivals, all arrivals are independent events, and independent events cannot be mutually exclusive, unless one of them is the null event.

If this were pointed out, the student might say, 'Of course! I should have multiplied the probabilities.' This would then imply the result that the probability of two arrivals in two hours is the square of the probability of one arrival in one hour. Unfortunately, this is also wrong, and gives a probability that is too small.

To show explicitly the relation between a number of arrivals in two hours and a number of arrivals in one hour, we may reason as follows. The event 'two arrivals in two hours' can be realised as one arrival in each of two consecutive hours, but it can also be obtained in two other ways – namely, two arrivals in the first hour and none in the second hour, or no arrivals in the first hour and two in the second hour. The probabilities of these three mutually exclusive events can now be summed to find the probability of two arrivals in two hours.

If you prefer to see the result algebraically:

$$\Pr(N2=2) = \Pr(N1=2).\Pr(N1=0) + \Pr(N1=1).\Pr(N1=1) \\ + \Pr(N1=0).\Pr(N1=2)$$

where $N1$ represents the number of arrivals in one hour (either the first hour or the second hour) and $N2$ represents the number of arrivals in two hours.

Evaluating first the left hand side of this equation, a probability from the Poisson distribution with parameter 2λ: $\Pr(N2=2)=\dfrac{1}{2}(2\lambda)^2 e^{-2\lambda}=2\lambda^2 e^{-2\lambda}$.

Next, the right hand side:

$$\Pr(N1=2).\Pr(N1=0)+\Pr(N1=1).\Pr(N1=1)$$
$$+\Pr(N1=0).\Pr(N1=2)=\frac{1}{2}\lambda^2 e^{-\lambda}e^{-\lambda}+\left(\lambda e^{-\lambda}\right)^2+e^{-\lambda}\frac{1}{2}\lambda^2 e^{-\lambda}=2\lambda^2 e^{-2\lambda}$$

Note that $P(N2 = 2)/P(N1 = 1) = 2\lambda e^{-\lambda}$, and this is always less than 2, whatever the value of λ.

References

Print

Schilling, M.F., Watkins, A.E. and Watkins, W. (2002). Is human height bimodal? *The American Statistician* **56**, 223–229.

Stigler, S.M. (1986). *The History of Statistics: The Measurement of Uncertainty Before 1900*. Harvard University Press.

Online

[13.1] Vaughan, T.S. (2008). In search of the memoryless property. In: Mason, S.J. *et al.* (eds.). *Proceedings of the 2008 Winter Simulation Conference*, Miami, USA, pp. 2572–2576. At http://www.informs-sim.org/wsc08papers/322.pdf

[13.2] Clarke, R.D. (1946). An application of the Poisson distribution. *Journal of the Institute of Actuaries* **72**, 481. Download at: http://www.actuaries.org.uk/research-and-resources/documents/application-poisson-distribution

Answers – Chapter 14

Question 14.1

a) The tangents intersect the z-axis at $z = -2$ and $+2$.

b) The ordinate at $z = 0.35958$ cuts off an upper tail area of (approximately) 0.35958. In other words, this value is the unique solution of the equation

$z = 1 - \Phi(z)$, where Φ is the (cumulative) distribution function of the standard normal density.

c) When $z = 6$, $\dfrac{1}{\sqrt{2\pi}}\exp\left(-\dfrac{1}{2}z^2\right) = 6.076 \times 10^{-9}$, approximately.

When $z = 0$ (at the mode), $\dfrac{1}{\sqrt{2\pi}}\exp\left(-\dfrac{1}{2}z^2\right) = 3.989 \times 10^{-1}$, approximately.

To find the scaled height of the curve at the mode, we must solve the proportionality problem $6.08 \times 10^{-9} : 1 \text{ mm} :: 3.99 \times 10^{-1} : x \text{ mm}$

We find $x = 65{,}659{,}969$ millimetres = 65.7 kilometres approximately.

Evidently, it is impossible to draw the standard normal distribution to this scale on paper as far as $z = 6$.

Question 14.2

Sheppard's tables of areas under the normal curve give, for example, the following values of the cumulative standard normal distribution: $P(Z < 2) = 0.9772499$; $P(Z < 4) = 0.9999683$; $P(Z < 6) = 0.9999999990$. The corresponding values from *WolframAlpha* (online at [14.5]) are: 0.97724987; 0.99996833; 0.99999999901. Sheppard's achievement is impressive, all the more so because his calculations were done by hand!

Sheppard wanted his results to be practically useful for mathematically interpolating cumulative probabilities for values of Z that he had not tabulated. His paper gives detailed formulae for this procedure, and notes that interpolation calculations unavoidably involve some loss of accuracy. He showed that his normal area tabulations to seven decimal places would enable calculation of interpolated values to an accuracy sufficient for all practical purposes.

It is interesting to note in this connection that there was, in the century 1850–1950, a competitive spirit in the air that was quite unmindful of practicality. This was prior to the era of the electronic computer – when a 'computer' was actually a person! At that time, there were many who celebrated computation to extremely high accuracy (e.g. of the value of π) as an achievement in itself. In his 1872 book, *A Budget of Paradoxes*, volume II, online at [14.6], the mathematician Augustus de Morgan commented: 'These tremendous stretches of calculation ... prove more than the capacity of this or that computer for labour and accuracy; they show that there is in the community an increase in skill and courage' (pages 63–64).

The achievements of human computers reached their zenith in the US in the so-called 'Mathematical Tables Project' (see online at [14.7]), that extended over the decade 1938–48. The many participants in this project evaluated a large number of mathematical functions to very high accuracy. These detailed tabulations were progressively published in 28 volumes, culminating in a hugely cited reference volume, *Handbook of Mathematical Functions*, which was compiled by Milton Abramowitz and Irene Stegun (1964), two veterans of the project.

Question 14.3

Galton delighted in his empirical confirmation that so many variables, plotted as frequency distributions of 'raw' measurements, closely follow the (theoretical) normal distribution, even in its very 'thin' tails. What such variables turn out to have in common is that their values can be interpreted as *small* random deviations from some fixed standard (or 'norm'). (Incidentally, some historians of statistics think that Galton settled on the term 'normal distribution' because of its connection to the idea of a norm in just this context.)

However, there are lots of variables that cannot be interpreted in this way – for instance, variables with highly skewed or bimodal empirical data distributions. The distribution of numbers of taxpayers by their taxable income is highly skewed in most countries. The distribution of numbers of drivers who die in car accidents by age is typically bimodal. Then there are symmetric and unimodal distributions that have 'fat' tails (compared with the normal distribution), meaning that the occurrence of large outliers is not improbable. The relative price changes of speculative shares listed on the Stock Exchange can (by definition) swing about wildly, even in the short run. A plotted distribution of the daily average of relative price changes of a set of speculative shares is usually fat-tailed.

For variables with a skewed distribution, mathematical transformations of their raw data to approximate normality are available. An example is found in the answer to QUESTION 14.4. However, there are no such transformations for bimodal or fat-tailed empirical distributions. They must be modelled by other kinds of statistical distributions – see CHAPTER 24 for examples.

Galton was also thrilled by the 'cosmic' dependability, as he thought of it, of the Central Limit Theorem (CLT) effect, whereby the mean of samples from seemingly any population distribution has a distribution that approximates the normal ever better as the sample size increases. The dependability of the CLT effect is certainly extensive; it applies to samples from all finite populations.

But it is different when one thinks in terms of a *model* for a finite population. If one chooses, as a model, one that is valid also for a (theoretically) infinite population, one discovers that there are some such models for which the

CLT fails. Thus, the CLT effect is not, in fact, universal. For example, the Cauchy distribution, which has an infinite range, is one choice of model for the distribution of a symmetric fat-tailed variable (see CHAPTER 24 for details). The distribution of the mean of samples from a Cauchy-distributed population is never better approximated by the normal distribution as the sample size increases. Instead, the mean is always exactly Cauchy-distributed.

So, was Galton wildly overstating his case? Given what a large proportion of practical situations in data analysis, across all disciplines, conform to Galton's celebration of a 'cosmic order', there is really little cause to quibble with his assessment. Without the wide applicability of the normal distribution and the CLT effect, statistics could not be the unified set of principles and techniques for analysing real-world data from almost any source that it is.

Question 14.4

In practical statistical work, what we call (approximately) 'normally distributed data' are data for which the histogram is symmetric and has a characteristic 'bell' shape. The symmetry is judged from a skewness measure of (close to) zero, and the 'bell' shape translates into a kurtosis measure, using the standard definition, of (close to) 3.

If the original data consist only of positive values, and show a mild positive skewness, then a logarithmic transformation may result in a distribution that is more like a normal. This is because taking logarithms 'shrinks' the larger positive values to a greater extent than the smaller positive values. If the original data are very highly skewed, a reciprocal transformation may achieve the desired normality. In both cases, the transformation will change not only the skewness, but also the kurtosis of the data distribution.

In general, it is difficult to decide in advance which of these transformations will produce a better approximation to normality but, with modern statistical packages, it is easy to try them both. Rather than trying to judge the approximation to normality from the histogram, it is easier to use a normal probability plot – a graphical tool that yields a straight line for perfectly normal data.

Question 14.5

The density function of the standard normal distribution is $f(x) = [1/\sqrt{(2\pi)}]$ $\exp(-\frac{1}{2}x^2)$ and that of the chi-squared distribution with 1 degree of freedom is $g(x) = [1/\sqrt{(2\pi)}] (1/\sqrt{x}) \exp(-\frac{1}{2}x)$. These two density functions are plotted in FIGURE 26.4 (note that the chi-squared distribution is defined only for $x > 0$). The plot shows that the two graphs intersect at $x = 1$.

We can prove that these two graphs have no other intersection points and, therefore, that at $x = 1$ the graphs are actually tangential.

The two graphs intersect where $f(x) = g(x)$, that is, where $\exp(-\frac{1}{2}x^2) = (1/\sqrt{x})\exp(-\frac{1}{2}x)$.

Taking natural logs of both sides and simplifying, we find $x^2 - x = \ln(x)$. This condition is satisfied when $x = 1$. FIGURE 26.5 shows that $x(x-1)$ is concave up, while $\ln(x)$ is concave down. Thus, there are no other intersection points of the two density functions, which implies that the density functions are tangential at $x = 1$.

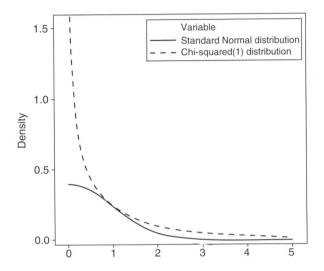

Figure 26.4 Density functions of $N(0,1)$ and chi-squared (1 df).

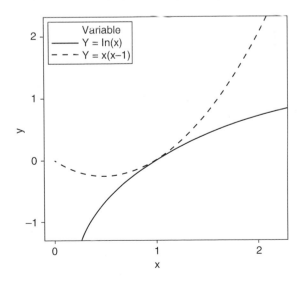

Figure 26.5 Graphs of $x(x-1)$ and $\ln(x)$ vs x.

References

Print

Abramowitz, M. and Stegun, I. (eds, 1964). *Handbook of Mathematical Functions with formulas, graphs, and mathematical tables.* National Bureau of Standards.

Online

[14.5] http://www.wolframalpha.com
[14.6] de Morgan, A. (1872). *A Budget of Paradoxes*, volume II. At http://www.gutenberg.org/ebooks/26408
[14.7] http://en.wikipedia.org/wiki/Mathematical_Tables_Project

Answers – Chapter 15

Question 15.1

It is important at the outset to define the sense in which an estimator is judged to be best. This example shows why.

It is certainly true that the sample mean (except in the case of a few out-of-the-ordinary theoretical probability distributions) is the minimum-variance unbiased (MVU) estimator of the population mean, and is in this sense the 'best' estimator. However, it is *not* true, in general, that the sample median is the MVU estimator of the population median. For example, in the case of the normal distribution, $N(\mu, \sigma^2)$, both the sample median and the sample mean are unbiased estimators of the population median (which is identical to the population mean). But, whereas the sample median has variance $(\pi/2)(\sigma^2/n)$ when the sample size, n, is large (see QUESTION 6.4), the sample mean has variance σ^2/n. Thus, the sample mean is more efficient here than the sample median for estimating the population median. In fact, the sample mean is the MVU estimator in this case, too, and is *in this sense* the 'best' estimator.

So, the best estimator of a particular population parameter is *not necessarily* the corresponding sample statistic.

Question 15.2

The correct answer is (approximately) 0.83, not the 'obvious' value 0.95.

The population distribution is normal $N(\mu, \sigma^2)$, with unknown mean μ and known variance σ^2. In the expression 'a 95% confidence interval for μ', the confidence coefficient 95% refers to the percentage of confidence intervals in a large number of replications that will, in theory, contain μ. The formula for the confidence interval depends directly on the distribution of the sample mean around the population mean. This distribution is $N(\mu, \sigma^2/n)$, where n is the sample size.

However, the probability that the mean (call it m_2) of a replicated sample will fall within a 95% confidence interval constructed around the mean (call it m_1) of the initial sample depends directly on the distribution of m_2 around m_1. Since the expectation of $(m_2 - m_1)$ is 0, the distribution of m_2 around m_1 is equivalent to the distribution of $(m_2 - m_1)$ around zero. Because the two samples are independent, the population variance of $(m_2 - m_1) = \text{var}(m_2) + \text{var}(m_1) = \sigma^2/n + \sigma^2/n = 2\sigma^2/n$. So $(m_2 - m_1)$ is distributed as $N(0, 2\sigma^2/n)$.

A little reflection should make it apparent that the probability that m_2 will lie within the 95% confidence interval for μ based on m_1 is the probability that $(m_2 - m_1)$ falls between $-1.96\sigma/\sqrt{n}$ and $+1.96\sigma/\sqrt{n}$ on $N(0, 2\sigma^2/n)$. That probability is $P(|z| < 1.96/\sqrt{2}) = P(|z| < 1.386)$, where z is a standard normal variable. Its value is approximately 0.83.

For an account of the importance of this result to statistical practice, see Cumming *et al.* (2004), and also Cumming (2006), online at [15.3].

Question 15.3

Each estimator is unbiased. However, the mean of the pooled data is more efficient – that is to say, its variance is lower unless the two sample sizes are equal, in which case the two estimators are the same. To prove this result on relative efficiency, confirm that the variance of the average of sample means is $\frac{1}{4}(\sigma^2/n_1 + \sigma^2/n_2)$, and that the variance of the mean of the pooled data is $\sigma^2/(n_1 + n_2)$. Then show that the former variance exceeds the latter unless $n_1 = n_2$.

Question 15.4

If a variable, X, is distributed as $N(\mu, \sigma^2)$, with μ assumed to be unknown, and $\bar{X}$ is the mean of a random sample of size n from this population, then $s^2 = \Sigma(X - \bar{X})^2 / (n-1)$ is the (unique) minimum-variance unbiased (MVU) estimator of σ^2. The formal proof of this result is by no means elementary. An accessible, though necessarily technical, account is given, for example, in Mood and Graybill (1963), pages 175–178, as well as in Roussas (1997), pages 284–292. In contrast to s^2, $\tilde{\sigma}^2 = \Sigma(X - \bar{X})^2 / (n+1)$ is a minimum mean squared error (MMSE) estimator of σ^2, and is evidently biased.

Clearly, these estimators, $\tilde{\sigma}^2$ and s^2, will produce very similar numerical values in practice, unless the sample size is quite small. In principle, then, a formal choice between them will be made on theoretical grounds. First, we shall explore these theoretical grounds, and then make the formal choice.

As mentioned in the answer to QUESTION 6.5, the mean square error (MSE) of an estimator is defined as the sum of the estimator's variance and the square of its bias. In symbols, if $\hat{\theta}$ is an estimator of some parameter, θ, of a distribution, based on a sample of a fixed size, then the MSE of $\hat{\theta}$ is defined as $E(\hat{\theta} - \theta)^2$, its variance as $E(\hat{\theta} - E\hat{\theta})^2$, and its bias as $(E\hat{\theta} - \theta)$. By writing the MSE in the form $E[(\hat{\theta} - E\hat{\theta}) + (E\hat{\theta} - \theta)]^2$, you can easily confirm that $E(\hat{\theta} - \theta)^2 = E(\hat{\theta} - E\hat{\theta})^2 + (E\hat{\theta} - \theta)^2$, that is, MSE = variance + (bias)^2.

As outlined in this chapter's Overview, once the MVU (i.e. the most efficient *unbiased*) estimator of a parameter has been identified, it is natural to ask whether there might be a *biased* estimator that is yet more efficient than the MVU estimator. Then, rather than being concerned about a perceived 'conflict' of criteria (between unbiasedness and efficiency) in choosing an estimator, one could accept some bias in an estimator, if it had an overcompensating increase in efficiency over the MVU. Why? Because the estimates generated by that estimator would, on average, lie closer to the parameter than the MVU estimator. Given the definition of the MSE, the method of MMSE estimation is an obvious path to discovering such an estimator *if there is one* (i.e. if the method doesn't fail).

Let's see where MMSE estimation leads in this normal distribution context. A logical beginning is to ask whether there is an estimator of the form cs^2 (where c is a positive constant), which (though biased) has a variance sufficiently smaller than that of s^2 as to make its MSE smaller than that of s^2 as well. The answer to QUESTION 6.5 gives the results $E(s^2) = \sigma^2$ and $\text{var}(s^2) = E(s^2 - \sigma^2)^2 = 2\sigma^4/(n-1)$. From these two results, we find $E(s^4) = \sigma^4[(n+1)/(n-1)]$. The MSE of cs^2 is $E(cs^2 - \sigma^2)^2$. Expanding, we get MSE $= c^2 E(s^4) - 2c\sigma^2 E(s^2) + \sigma^4$. To find the value of c which yields the minimum of this MSE, we differentiate with respect to c, and set the result to zero: $2cE(s^4) - 2\sigma^2 E(s^2) = 0$. Substituting for the expected values, we find $c = [(n-1)/(n+1)]$. Thus, the MMSE estimator of σ^2 here is $\tilde{\sigma}^2 = \sum(X - \bar{X})^2/(n+1)$.

There are two things worth noting about $\tilde{\sigma}^2$:

a) Its bias is $-2\sigma^2/(n+1)$, which diminishes as the sample size increases.
b) Its MSE is $2\sigma^4/(n+1)$, which is smaller than the MSE (= variance) of s^2, i.e. $2\sigma^4/(n-1)$. So, despite its bias, $\tilde{\sigma}^2$ is, on average, closer to σ^2 than is s^2. The relative MSE advantage that $\tilde{\sigma}^2$ has over s^2 is $(n-1)/(n+1)$. This is greatest when n is small.

What should now be our formal conclusion in the comparison between $\tilde{\sigma}^2$ and s^2? $\tilde{\sigma}^2$ is a little biased, but it is *so much more efficient* than s^2 that the MSE of $\tilde{\sigma}^2$ is smaller than that of s^2. On the MSE criterion, $\tilde{\sigma}^2$ is the preferred estimator.

Question 15.5

a) We follow the same procedure as we did in the answer to QUESTION 15.4 when estimating σ^2. Denote by $\tilde{X}$ the estimator of μ that is of the form $c\bar{X}$. Then the bias of $\tilde{X}$ is $E(c\bar{X} - \mu) = \mu(c - 1)$ and the variance of $\tilde{X}$ is $c^2 \operatorname{var}\bar{X} = c^2\sigma^2/n$. Thus, the MSE of $\tilde{X} = c^2\sigma^2/n + [\mu(c - 1)]^2 = c^2[\mu^2 + \sigma^2/n] - 2c\mu^2 + \mu^2$. Differentiating with respect to c and equating the result to zero yields the value of c that minimises the MSE, namely,

$$c = \left[\frac{\mu^2}{\mu^2 + \dfrac{\sigma^2}{n}} \right]. \text{ So, } \tilde{X} = \left[\frac{\mu^2}{\mu^2 + \dfrac{\sigma^2}{n}} \right] \bar{X}.$$

Because $\tilde{X}$ is a function of μ, the parameter to be estimated, $\tilde{X}$ is useless as an estimator of μ. This is the sign that the method of MMSE estimation has failed.

b) James and Stein's counterintuitive result arises in the context of estimating the population means of several independently-distributed normal variables, when the population variances are known and are all equal and when we have a sample of data on each one of the variables. We shall approach their result from what we assume is already familiar territory for you.

When an estimator is needed for the mean of a single normal distribution with known variance, all of the standard criteria of a good estimator point to the sample mean. That is because, in this context, the sample mean is the MVU estimator. It is also the maximum likelihood estimator.

If the criterion of minimum mean square error (MMSE) is added (so as to admit consideration of the possibility that a biased estimator may have a yet smaller variance), the superiority of the sample mean is unaffected since (as the answer to QUESTION 15.5(a) shows) MMSE estimation fails to provide an estimator in this context. So, there is no biased estimator of the population mean that improves on the (unbiased) sample mean. Expressing this in technical language: the sample mean is here an 'admissible' estimator in terms of MSE.

If, now, this estimation problem is generalised to that of estimating the population means of several independently distributed normal variables, when the population variances are known and are all equal, it seems intuitively clear that each sample mean will be an admissible estimator (in terms of MSE) of its corresponding population mean.

James and Stein, however, proved a powerfully counterintuitive result – see James and Stein (1961), online at [15.4]. They showed that the sample mean is admissible when there are one or two means to be estimated. However, in the case of three or more means, the sample mean is inadmissible! We alert you that to follow this paper (and the papers cited by those authors) in detail requires advanced statistical knowledge.

The authors also presented a biased estimator, now unsurprisingly called the James-Stein estimator, of each mean. For a set of more than two means, this estimator has the property that the sum of the mean square errors of its estimates is less than the sum of the mean square errors (= variances) of the individual (unbiased) sample means.

In symbols, suppose we have n observations on each of k independently distributed normal variables, $X_{i,j}$ ($i = 1, 2, \ldots k; j = 1, 2, \ldots n$), the variables having distinct means, μ_i, but a common variance, σ^2 (assumed known). The sample mean for each variable, $\bar{X}_i = \sum_{j=1}^{n} X_{ij}$. The means $\bar{X}_i$ ($i = 1, 2, \ldots k$) are independently distributed, each being $N(\mu_i, \sigma^2/n)$.

The James-Stein estimator of μ_i is, then, $\tilde{X}_i = \left[1 - \dfrac{(k-2)\sigma^2}{\sum_{1}^{k} \bar{X}_i^2} \right] \bar{X}_i$, for

$k \geq 2$. If $k > 2$, *the sum of* the MSEs of the k estimators $\tilde{X}_i$ is shown to be less than $k\sigma^2/n$, which is *the sum of* the variances of the k estimators $\bar{X}_i$.

Note the italics in the previous sentence: the optimality criterion under which the James-Stein estimator is derived is minimising the sum of the MSEs of the k estimators $\tilde{X}_i$. This does not imply that each $\tilde{X}_i$ has a lower MSE than the corresponding $\bar{X}_i$. Some $\tilde{X}_i$ will have a lower MSE and some a higher MSE. Nor can one determine exactly which of the μ_i has been better estimated by the James-Stein estimator. What is established is that, *on average*, the μ_i are better estimated (in terms of MSE) by using the James-Stein estimator, rather than the sample mean.

The widespread interest which the James and Stein paper created among statisticians was not solely because of its potential, in practice, to yield an estimate with much reduced MSE. Two theoretical paradoxes of the James-Stein estimator, in the context in which it first appeared, were far more responsible.

The first of these paradoxes is the finding that the sample mean is admissible in this context if $k = 1$ or $k = 2$, but not if $k > 2$. Why should optimal accuracy of estimation call for the James-Stein estimator when three or more independent means are simultaneously estimated, but not when only one mean or two independent means are estimated?

This paradox made the James-Stein estimator as astonishing to 20th century statisticians as the Central Limit Theorem (discovered by Laplace, see CHAPTER 12) was to their 19th century predecessors.

The second paradox is embedded in the estimator formula. Each $\tilde{X}_i$ is seen to depend (mathematically) not only on the corresponding sample mean $\overline{X}_i$ but also, through the element $\sum_1^k \overline{X}_i^2$, on the sample means of all the other (statistically) independent variables. This is puzzling indeed! Why should optimal accuracy of estimation of a particular population mean depend, in part, on the behaviour of a completely unrelated variable?

The first paradox is clarified heuristically in a creative way by Stigler (1990), online at [15.5]. The source of the second paradox is, rather obviously, that the optimality criterion, under which the James-Stein estimator is derived, is minimising *the sum of* the MSEs of the k estimators $\tilde{X}_i$. However, that simply prompts the question, why would one choose to estimate jointly the means of a set of *independent* variables? With empirical examples, Efron (1975) responds insightfully to this question.

A non-technical account of James-Stein estimation, with some perspectives beyond those we have presented here, is given by Efron and Morris (1977).

Since 1961, hundreds of articles have appeared on James-Stein estimation, some seeking to generalise the approach to cases where the initial assumptions (normality, independence, known and equal population variances, and use of the sum of MSEs as the optimality criterion) are varied, and others extending James-Stein estimation to wider contexts, including regression modelling. Similarities between James-Stein estimation and Bayesian inference have also been extensively investigated.

There is a small, but cogently argued, literature of 'dissent', which argues that the claims made for the theoretical superiority of James-Stein estimation are, in certain contexts, philosophically shaky and/or quantitatively exaggerated. It is well acknowledged, even by advocates of the James-Stein approach, that this is not a tool to be applied mechanistically – there are many traps for the unwary. Perhaps that is why James-Stein estimation does not seem, to us, to have revolutionised statistical practice in non-academic settings.

References

Print

Cumming, G., Williams, J. and Fidler, F. (2004). Replication and researchers' understanding of confidence intervals and standard error bars. *Understanding Statistics* **3**, 299–311.

Efron, B. (1975). Biased versus unbiased estimation. *Advances in Mathematics* **16**, 259–277.

Efron, B. and Morris, C. (1977). Stein's paradox in statistics. *Scientific American* **236**(5), 119–127.

Mood, A and Graybill, F. (1963). *Introduction to the Theory of Statistics*, 2nd edition. McGraw-Hill.

Roussas, G. (1997). *A Course in Mathematical Statistics*. Academic Press.

Online

[15.3] Cumming, G. (2006). Understanding replication: confidence intervals, p values, and what's likely to happen next time. *Proceedings of the Seventh International Conference on Teaching Statistics* (ICOTS7). At: http://www.stat.auckland.ac.nz/~iase/publications/17/7D3_CUMM.pdf

[15.4] James, W. and Stein, C. (1961). Estimation with quadratic loss. *Proceedings of the Fourth Berkeley Symposium on Mathematical Statistics and Probability* **1**, 361–379. At: http://projecteuclid.org/euclid.bsmsp/1200512173

[15.5] Stigler, S.M. (1990). A Galtonian perspective on shrinkage estimators. *Statistical Science* **5**, 147–155. At http://projecteuclid.org/euclid.ss/1177012274

Answers – Chapter 16

Question 16.1

It was initially R.A. Fisher, in 1925, who gave the 5% significance level special weight, though later he adopted a more flexible view – urging that the context of the hypothesis test be considered in choosing a significance level. In fact, it was Neyman and Pearson who subsequently insisted on limiting the choice of significance levels to a standard set, and on treating the corresponding significance points as if they objectively demarcated rejection from acceptance of the null hypothesis.

In the chapter 'Why $P = 0.05$?' in his book, *The Little Handbook of Statistical Practice* (online at [16.3]), Dallal quotes from Fisher's 1925 book

Statistical Methods for Research Workers: 'The value for which $P = 0.05$, or 1 in 20, is 1.96 or nearly 2; it is convenient to take this point as a limit in judging whether a deviation ought to be considered significant or not. Deviations exceeding twice the standard deviation are thus formally regarded as significant.'

Later writers have suggested that Fisher's rather dogmatic choice, however firmly he stated it in 1925, was essentially arbitrary. This is convincingly contradicted by Cowles and Davis (1982). They show that reasoning similar to Fisher's is traceable back to de Moivre and Gauss in the early 19th century.

Question 16.2

We assume the confidence interval (CI) for μ is two-sided and equal-tailed. If the value μ_0 is outside the 95% CI, then the null hypothesis $\mu = \mu_0$ will be rejected by the test at the 5% level of significance. This can be shown in a straightforward way from the formulae for the confidence interval and the rejection region of the test. Alternatively, here is a heuristic argument: the CI contains all the values of μ that are 'reasonable' at the 95% level, given the value obtained for the sample mean, and excludes those values that are 'not reasonable' – that is, the most extreme 5% of values. So, if the particular value μ_0 is not in the 95% CI, then the hypothesis that $\mu = \mu_0$ should be rejected as 'unreasonable', with a risk of decision error (that is, a level of significance) of 5%. In this way, the equal-tailed 95% CI corresponds to the two-sided test at the 5% level of significance. If we wanted to use the equal-tailed 95% CI to carry out a test of $\mu = \mu_0$ against a one-sided alternative, the procedure in this case would imply a 2.5% level of significance.

Though hypothesis tests and confidence intervals are *analytically* equivalent, subject to the interpretational refinements just given, the Overview explains that there are circumstances where the CI has greater *practical* utility than the equivalent test. In these circumstances, the CI is clearly preferable.

Question 16.3

The power curve for the equal-tailed test of H_0: $\mu = \mu_0$ against H_1: $\mu \neq \mu_0$, based on the mean of a random sample drawn from a normal population, $N(\mu, \sigma^2)$, where the value of σ^2 is known, typically has the shape of the solid line in FIGURE 26.6. The level of significance (corresponding to the ordinate at the minimum value) is 0.05, or 5%.

This figure is occasionally shown in statistics textbooks, sometimes with the comment that is resembles an upside-down normal distribution.

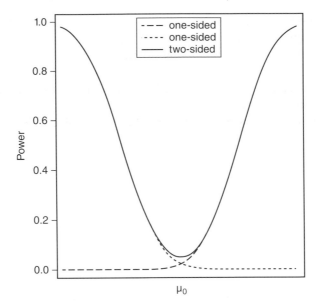

Figure 26.6 Power curves for two-sided (solid line) and one-sided (dotted lines) hypothesis tests of a normal mean.

It is, however, *not* an upside-down normal distribution. Accurately expressed, it is the sum of the ordinates of two S-shaped curves – a *cumulative normal* distribution and a *decumulative normal* distribution (i.e. a laterally reversed cumulative normal distribution) – that cross at the point where $\mu = \mu_0$, and where each has an ordinate value of 0.025.

The cumulative normal distribution is the power curve of the one-sided test of H_0: $\mu = \mu_0$ against H_1: $\mu > \mu_0$ at the 2.5% level of significance. The decumulative normal distribution is the power curve of the one-sided test of H_0: $\mu = \mu_0$ against H_1: $\mu < \mu_0$ at the 2.5% level of significance. The power curve for the two-sided test is the sum of these, in the same way as the probability of a type I error in the two-sided test is the sum of the probabilities of a type I error in each of the one-sided tests.

A graphical alternative to this verbal explanation is found in Shoesmith (1983). Shoesmith shows that, when the power curve for the two-sided test is plotted on normal probability paper, it looks like a blunted V – two diagonal lines, joined by a curve around the null hypothesis value $\mu = \mu_0$.

Question 16.4

Denote the null hypothesis by H_0 and the alternative hypothesis by H_1. It should be made clear, first of all, that the level of significance is not simply the probability that H_0 will be rejected. Rather, it is the probability that H_0 will be rejected (erroneously), *given that H_0 is actually true.* Thus, 0.05 = P(H_0 rejected | H_0 true). The other piece of information we have is the power of this test, the power being the probability that H_0 will be rejected (correctly), *given that H_0 is actually false.* Thus, 0.90 = P(H_0 rejected | H_0 false) = P(H_0 rejected | H_1 true)

The question is now expressible as 'what is the value of P(H_0 true | H_0 rejected)?'. This can be found by means of Bayes' Theorem (see CHAPTER 20 for some background on this theorem):

$$P\left(H_0 \text{true}|H_0\text{rejected}\right)=$$

$$\frac{P\left(H_0 \text{ rejected}|H_0 \text{ true}\right).P\left(H_0 \text{ true}\right)}{\left[P\left(H_0 \text{ rejected}|H_0 \text{ true}\right).P\left(H_0 \text{ true}\right)\right]+\left[P\left(H_0 \text{ rejected}|H_1 \text{ true}\right).P\left(H_1 \text{ true}\right)\right]}$$

We have the values of the conditional probabilities in the expression on the right hand side of this relation. But the expression can be fully evaluated only if values for P(H_0 true) and P(H_1 true) are known. All we can say about the probability that H_0 is true is that this probability is, in general, smaller when we incorporate the information that the test has actually rejected H_0 – as we do when we evaluate P(H_0 true | H_0 rejected) – than it would be without incorporating that information.

Thus, our conclusion must be that the question is unanswerable numerically without further information, as indicated.

Question 16.5

Firstly, we can assume that only rolls of 9 and of 10 with the three dice will give any information on the relative probabilities of 9 and 10. Since Pr(9) = 25/216 and Pr(10) = 27/216, Pr(9 or 10) = 52/216, so only that proportion of observations on average will be useful for our purpose (one could, alternatively, use a symmetry argument and include also rolls of 11 and of 12).

Secondly, we shall interpret the question as an empirical test of the null hypothesis that Pr(10) = Pr(9), against the alternative that Pr(10) > Pr(9) – though one could argue for a two-sided alternative. Putting Pr(10)/Pr(9 or 10) = π, we write H_0: π = 0.5 and H_1: π > 0.5. Given that the true value of π = 27/52 = 0.51923, and that we set the level of significance at 0.05, we will need 5786 sample observations on π to obtain a power of 0.90 in this test. This number can be obtained by direct calculation (the normal approximation to the binomial is excellent here) or, more easily, from Russ Lenth's power and sample size website at [16.4] (or by using Minitab's 'power and sample size' option, if you have access to that statistical package). This means that we would expect to require 5786 × 216/52 = 24,304 observations of the roll of three dice to establish our result (or half that many – 12,152 – if the symmetry argument were used).

Could anyone really have noticed such a tiny difference of 2/216 (=0.0093) between the chance of a 9 and a 10? Galileo implies that someone had! It suggests that inveterate gamblers of that era acquired a quite uncanny perceptiveness about chances from their experience of tens of thousands of rolls. The source of this perceptiveness is all the more intriguing, since such gamblers are hardly likely to have kept meticulous written records of the fall of the dice.

Note: this question was originally posed in Hald (2003), page 41.

References

Print

Cowles, M. and Davis, C. (1982). On the origins of the .05 level of statistical significance, *American Psychologist* **37**, 553–558.

Hald, A. (2003). *A History of Probability and Statistics and their Applications before 1750*. Wiley.

Shoesmith, E. (1983). Simple power curve constructions. *Teaching Statistics* **5**, 78–83.

Online

[16.3] http://www.jerrydallal.com/LHSP/LHSP.htm

[16.4] http://www.math.uiowa.edu/~rlenth/Power/

Answers – Chapter 17

Question 17.1

The mythical animal is the unicorn, and its diurnal metabolic activity was 'investigated' in Cole (1957). A set of random data was transformed by moving averages and other seemingly reasonable techniques until, lo and behold, a cycle appeared. This cycle the author interprets, deadpan, in these words: 'Eliminating the effect of the lunar periodicity shows that the peak of endogenous activity occurs at "3 a.m." and that the minimum occurs exactly

12 hours later. The unicorn obviously tends to be active in the early morning and quiescent in midday. The "midmorning" dip in activity indicated in the figure remains unexplained but may possibly be a subject for future research.'

A more recent study, similarly provocative but in a different analytical setting, was carried out by Bennett *et al.* (2009), whose research report poster is online at [17.5]. It concerned an 'investigation' of the neural activity of a dead salmon, as measured by functional magnetic resonance imaging. According to the results, the salmon appeared to be responsive to the psychological task it was set, but this was only the case when the data were analysed without any correction for multiple testing. When such a correction was applied, the salmon (not surprisingly!) showed no significant results. A brief perspective on the impact this study has had on the field of imaging in neuroscience appeared in *Nature* and is online at [17.6].

Both of these whimsical studies have a serious message – without vigilant scrutiny of the meaningfulness of the data and the validity of the statistical procedures used, data snooping can readily produce plausible conclusions that are actually nonsense.

Question 17.2

In the context of cyber security, data snooping means collecting online information about individuals from supposedly secure archives, either by transgressing limited access rights or by illicit hacking. The term is particularly used in situations where disparate sources of data on an individual are merged by the snooper, often for commercial gain. The term shares with the statistical use of the term the implication that something improper is being done. Professional statisticians would usually have ethical concerns about: (a) collecting personal information; (b) matching up different sources of personal information; and (c) using personal information for financial gain. Before doing (a) or (b), they would expect to specify an ethics protocol regarding confidentiality and informed consent, and they would usually avoid (c).

Question 17.3

For a single hypothesis test where the null hypothesis is actually true, call the probability of rejecting the null hypothesis α. This defines the level of significance of the test and equals the size of the type I decision error.

Now consider two *independent* tests. In the case where each null hypothesis is true, the probability of rejecting a null hypothesis at least once is $1 - (1 - \alpha)^2 = 2\alpha - \alpha^2$. For small values of α (such as the commonly used levels of significance $\alpha = 0.05$ or 0.01), $2\alpha - \alpha^2$ is very slightly less than 2α, and so is well approximated by 2α. This approximation is the logical basis of the Bonferroni adjustment. If we then apply the Bonferroni adjustment, and carry out each individual test using a significance level of $\alpha/2$, the overall level of significance (and, hence, the size of the type I error) of the 'family' of two tests will be at most α.

If the two tests are, for instance, performed using the same data, they will be *dependent* tests, in the sense that the events 'reject null hypothesis 1' and 'reject null hypothesis 2' are dependent events.

Assume, as above, that each null hypothesis is true, and that each test is carried out with level of significance α. Then, on a Venn diagram, the two events just mentioned will be represented by circles with an overlap corresponding to a joint probability *not* of α^2 (which is the case for independent tests), but of $\alpha^2 - \theta$, where θ is a value that lies between 0 (where the circles do not overlap at all, so both null hypotheses cannot be rejected at the same time) and α (where the two circles coincide, so rejection of one null hypothesis implies rejection of the other). Correspondingly, the probability of rejecting a null hypothesis at least once will vary between $1 - (1 - \alpha)^2 = 2\alpha - \alpha^2$ and $1 - (1 - \alpha) = \alpha$. So, the overall level of significance for both tests is still less than 2α. For small values of α, these limits are approximately 2α and α. Thus here, too, carrying out each test at a significance level of $\alpha/2$ will ensure that the overall significance level is, at most, α. Martin Bland, (online at [17.7]), provides a useful complementary discussion with examples.

Question 17.4

The null hypothesis is that Paul has no psychic ability or, equivalently, that he has an equal chance of choosing the winning or the losing team. (The number of times he chooses the winning team has a binomial distribution, with $\pi = 0.5$.) The alternative hypothesis is that Paul has psychic ability – that is, he has a greater than even chance of choosing the winning team. (Assuming a one-sided test, the number of times he chooses the winning team has a binomial distribution, with $\pi > 0.5$.) Since Paul chose eight out of eight correctly, the p-value for the test is $0.5^8 = 0.004$. This gives evidence for Paul's psychic ability at better than the conventional 0.05 level for a statistical test. However, there may have been other reasons for Paul's success that are not due to psychic ability. For instance, Paul's favourite food may have been placed – by chance or by design – only in the box corresponding to the team that was more likely to win.

If we believe – as seems reasonable – that octopuses are generally unable to predict the results of soccer matches, how do we explain the surprise of

Paul's success? Maybe it is simply due to selective reporting. There were probably many thousands of attempts to predict all the results of the 2010 World Cup. Some of these may have been made using animals (including other octopuses). Many of the people, and most of the animals, would have had limited success, but a few – by chance or (human) skill – will have ended up with a complete set of correct predictions. These few (including Paul) became the only ones to be very widely reported. Were they, however, to be viewed against the huge number of attempts *at the outset* to predict all the results accurately, this small number of completely correct predictions would be seen as being compatible with the notion that they were achieved just by a run of good luck.

Selecting one correct prediction from an unmentioned large number of (less successful) predictions is akin to data snooping, in the sense that snooping describes the situation where multiple statistical procedures are performed on a set of data, but only the 'best' result that achieves statistical significance is reported.

We note that selective quoting of optimal results is difficult for others to detect, unless there is a commitment by published authors to make *all* their empirical results (as well as their data) publicly accessible – and not only the results that appear in their published work. In other words, studies that showed no significant results, or that were, in some other way, not of interest to academic referees or journal editors, should also be preserved – perhaps on a dedicated online database. These actions are, in fact, expected of all scientists who adhere to the principles of the modern Open Science movement (see online at [17.8]). The purpose goes far beyond detecting selective quotation, of course. Every competently produced empirical finding has something informative to offer future researchers seeking a widely comprehensive view.

Question 17.5

The statisticians were Yoav Benjamini and Yosef Hochberg, both of Tel Aviv University. In Benjamini and Hochberg (1995), they proposed a procedure that limited the proportion of 'significant' results that were, in fact, due to chance. The 'false discovery rate' (or FDR) is controlled using a sequential procedure that depends on the number of tests carried out. The significance level of each individual test is reduced, but not as much as when a Bonferroni adjustment is used. The result is an ability to make statements such as 'of the 100 rejected null hypotheses, at most 5% are falsely rejected (though we don't know which ones they are)'. An article by McDonald, online at [17.9], gives a good discussion of the principles.

Incidentally, the authors of the dead salmon study, referred to in the answer to QUESTION 17.1, referenced their use of the Benjamini-Hochberg correction for multiple testing to demonstrate the truth that dead salmon don't respond to social cues.

References

Print

Benjamini, Y. and Hochberg, Y. (1995). Controlling the false discovery rate: a practical and powerful approach to multiple testing. *Journal of the Royal Statistical Society, Series B* **57**, 289–300.

Cole, L.C. (1957). Biological clock in the unicorn. *Science* **125**, 874–876.

Online

[17.5] Bennett, C.M., Baird, A.A., Miller, M.B. and Wolford, G.L. (2009), Neural correlates of interspecies perspective taking in the post-mortem Atlantic Salmon: an argument for multiple comparisons correction. At: http://prefrontal.org/files/posters/Bennett-Salmon-2009.pdf

[17.6] Gewin, V. (2012). Turning point: Craig Bennett, *Nature*, **490**, no. 7420, 18 October, page 437. At: http://www.nature.com/naturejobs/science/articles/10.1038/nj7420-437a

[17.7] http://www-users.york.ac.uk/~mb55/intro/bonf.htm. This is an extract from the textbook, Bland, J.M. (2000). *An Introduction to Medical Statistics*. Oxford University Press.

[17.8] https://en.wikipedia.org/wiki/Open_science

[17.9] McDonald, J.H. (2014). Multiple comparisons, in *Handbook of Biological Statistics*. At: http://www.biostathandbook.com/multiplecomparisons.html

Answers – Chapter 18

Question 18.1

If we denote the original equation by Kg = $a + b$ Inch, and then let Inch = 2.54 cm, we can see that the new equation will be Kg = $\alpha + \beta$ Cm, where $\alpha = a$ and $\beta = 2.54b$. Thus, we can see that the intercept is the same in both regressions, but the slope is different. The values of r^2 will also be the same, since r^2 is calculated using standardised data, which have no units of measurement.

Question 18.2

The given data points all lie on the horizontal line $y = 3$. If you attempt to calculate r^2, you will find that both the numerator and the denominator expressions are zero. The value of the ratio $0/0$ is undefined. In other words, the coefficient of determination is, here, undefined. You can come to this conclusion without any calculation if you view r^2 from another perspective. The coefficient of determination measures strength of linear association between two variables. In the given dataset, X is a variable but Y is a constant. Hence, the coefficient of determination is here an invalid measure.

Question 18.3

Many women of the Khoikhoi people (called 'Hottentot' by early European colonisers) were observed to have steatopygous (i.e. protruding) buttocks, which seemed to be a source of fascination for Europeans in the 19th century. Galton, apparently fascinated himself, measured the contours of local women in a way which permitted him to remain at a proper 'Victorian' distance during the process! In a letter dated 23 February 1851 to his older brother, Darwin Galton, he explained effusively how he did it:

> 'I am sure you will be curious to learn whether the Hottentot Ladies are really endowed with that shape which European milliners so vainly attempt to imitate. They *are* so, it is a fact, Darwin. I have seen figures that would drive the females of our native land desperate… [A]s a scientific man and as a lover of the beautiful I have dexterously, even without the knowledge of the parties concerned, resorted to actual measurement… I sat at a distance with my sextant, and as the ladies turned themselves about, as women always do, to be admired, I surveyed them in every way and subsequently measured the distance of the spot where they stood – worked out and tabulated the results at my leisure.' (Reprinted in Pearson (1914), pages 231–232.)
> *Note*: milliners apparently also made bustles in the 19th century!

Question 18.4

The book is *The Biographer's Tale*, published in 2000 by the British author A.S. Byatt, who may be better known for her earlier novel, *Possession*. The character identified as CL is the Swedish botanist, Carl Linnaeus, today regarded as the founder of scientific taxonomy. The person identified as HI is the Norwegian playwright, Henrik Ibsen. Byatt's book is a wonderfully intriguing and erudite mix of fact and fantasy. You can read an interesting review (and a collection of quotes from other reviews) online at [18.3].

Question 18.5

Galton's diagram has child height (the 'explained' variable) on the horizontal axis, and mid-parental height (the 'explanatory' variable) on the vertical axis. If, as is nowadays conventional, we swap the axes, so that the 'explained' variable is on the vertical axis, then we must show that the regression line of child height on mid-parental height corresponds to the line joining the origin to the point where the tangent to the ellipse is *vertical*. Let X = child height and Y = mid-parental height. It will simplify the algebra of our demonstration, with no loss of generality, if we first transform X and Y into standardised variables, x and y, in each case by subtracting its mean and dividing by its standard deviation.

It is then reasonable, following Galton, to represent the population joint-distribution of x and y as a standard bivariate normal distribution with correlation coefficient ρ – that is, $(x, y) \sim N(0, 0, 1, 1, \rho)$. Standard formulae (see, for instance, page 6 online at [18.4]) give the equation of the elliptical contours as $x^2 - 2\rho xy + y^2 = c$ (c is a constant). The tangent to a contour ellipse will be vertical where dy/dx is infinite (i.e. $dx/dy = 0$).

Applying d/dy to the equation of the ellipse (using implicit differentiation where needed) gives $2x(dx/dy) - 2\rho(x + y\, dx/dy) + 2y = 0$. Setting dx/dy to zero results in $-2\rho x + 2y = 0$, that is, $y = \rho x$ for the locus of intersection points of the elliptical contours of the standard bivariate normal and their respective vertical tangents. The line $y = \rho x$ will be recognised as the equation of the *population* regression line in the theory of simple regression analysis of y on x.

Applying the method of least squares to data on x and y will evaluate the *sample* regression line of y on x, $y = rx$, where r is none other than the sample correlation coefficient between x and y and, evidently, an estimator of ρ.

References

Print

Pearson, K. (1914). *The Life, Letters and Labours of Francis Galton*, volume 1. Cambridge University Press.

Online

[18.3] http://www.complete-review.com/reviews/byattas/bstale.htm

[18.4] http://www.athenasc.com/Bivariate-Normal.pdf This is an excerpt from Bertsekas, D. and Tsitsiklis, J. (2002). *Introduction to Probability*, 1st edition. Athena Scientific.

Answers – Chapter 19

Question 19.1

The 'lady tasting tea', whose confident assertion prompted R.A. Fisher to consider how to incorporate the design of experiments into his developing framework of techniques of statistical inference, was Dr Muriel Bristol-Roach. Fisher's biographer, his daughter, describes the historic teatime at the Rothamsted Agricultural Research Station in Hertfordshire, apparently in 1921 or 1922:

> 'It happened one afternoon when [Fisher] drew a cup of tea from the urn and offered it to the lady beside him, Dr. B. Muriel Bristol, an algologist. She declined it, stating that she preferred a cup into which the milk had been poured first. "Nonsense," returned Fisher, smiling, "Surely it makes no difference." But she maintained, with emphasis, that of course it did. From just behind, a voice suggested, "Let's test her." It was William Roach who was not long afterward to marry Miss Bristol. Immediately, they embarked on the preliminaries of the experiment, Roach assisting with the cups and exulting that Miss Bristol divined correctly more than enough of those cups into which tea had been poured first to prove her case.
>
> 'Miss Bristol's personal triumph was never recorded, and perhaps Fisher was not satisfied at that moment with the extempore experimental procedure. One can be sure, however, that even as he conceived and carried out the experiment beside the trestle table … he was thinking through the questions it raised: How many cups should be used in the test? Should they be paired? In what order should the cups be presented? What should be done about chance variations in the temperature, sweetness, and so on? What conclusion could be drawn from a perfect score or from one with one or more errors?' (from Box (1978), page 134).

For further insights on the people involved and the rich statistical consequences of the occasion, see Lindley (1993) and Senn (2012).

Question 19.2

The Rothamsted Agricultural Research Station, called Rothamsted Research since 2002, lies outside the town of Harpenden in Hertfordshire, England. It was founded in 1843 on the extensive grounds of Rothamsted Manor.

The website of Rothamsted Research, online at [19.1], says about the institute's origins: 'The Applied Statistics Group continues to build on the strong tradition of statistical contributions to the research at Rothamsted, which started with the appointment of Ronald A. Fisher in 1919 and continued with many other distinguished statisticians, including Frank Yates, John Nelder, John Gower and Robin Thompson.' To this list we can add William Cochran, Oscar Irwin and John Wishart.

Indeed, Fisher founded the Statistical Laboratory at Rothamsted and worked there until 1933. It was there that research questions he faced in his empirical work led him to many of his towering theoretical contributions to statistical inference. His immediate access to large quantities of real-world data from agricultural field experiments at Rothamsted – especially from the so-called 'Classical Experiments' (in progress since the 1850s) – was helpful to him in trialling his new techniques in practice.

These techniques include the analysis of variance, the method of maximum likelihood estimation, the randomisation test, and the theory of experimental design. On the last of these, see his daughter's account in Box (1980). Fisher's extensive involvement with agricultural data generated a series of papers with the general title 'Studies in crop variation', and culminated in the publication of his practical manual Fisher (1925), as well as the theoretical text Fisher (1935). Both books have subsequently had multiple editions.

On leaving Rothamsted for a professorship at University College, London, Fisher summed up his fifteen years' work at Rothamsted in a chapter of its 1933 Annual Report. This is entitled, 'The contributions of Rothamsted to the development of the science of statistics', and is available online at [19.2].

Question 19.3

For every value of n, it is possible to construct an $n \times n$ square array of cells in which n different symbols are arranged so that each symbol appears only once in each row and in each column. If the symbols are letters of the Latin (today, we say 'Roman') alphabet, then the array is known as a Latin square. For $n = 3$, there are six different Latin squares, for $n = 4$ there are 576, and so on. Some of the mathematical properties of Latin squares up to order $n = 6$ were investigated in 1782 by the Swiss mathematician Leonhard Euler (1707–1783). R.A. Fisher introduced Latin squares into the design of statistical experiments in 1934, using letters to symbolise experimental treatments, so that each treatment is allocated just once to each row and just once to each column.

In the agricultural context of this question, suppose there are four treatments – that is, four different levels of fertiliser (in gm/square metre) applied to plots of ground. Call these levels a, b, c, d. A Latin square

Depth (in cm) of sowing the seed →		1	3	5	7
Natural soil moisture gradient of the land ↓	5%	a	b	c	d
	20%	c	d	a	b
	35%	d	c	b	a
	50%	b	a	d	c

Figure 26.7 A Latin square design.

corresponding to these four letters is shown in FIGURE 26.7. The corresponding 16 cells of the figure may be thought of as plots of ground.

The nuisance variables are soil moisture and depth of sowing. To use the Latin square design, the number of levels of each of the two nuisance variables must equal the number of levels of treatment in the factor of chief interest (here, the levels of fertiliser applied to the plots).

Across the columns for each row is one of four levels of soil moisture (measured as the percentage of a fixed volume of soil that is water): 5%, 20%, 35%, 50% (you could imagine this steady increase as reflecting a moisture gradient on a downhill slope). Across the rows for each column is one of the depths of sowing (measured in cm): 1, 3, 5, 7.

The Latin square design effectively controls the influence of soil moisture on crop yield, because each level of fertiliser is applied once at every level of soil moisture. This enables the influence of soil moisture to be computationally 'averaged out' of the multivariable relation linking crop yield to fertiliser, soil moisture, and depth of planting. The same reasoning applies to this design's control of the influence of depth of planting on crop yield.

Fisher thought that the Latin square used in any particular experiment should be chosen at random from all the squares of the appropriate order, though he did not have an objective way of doing so. Today, a common method is simply to pick a square and permute its rows or columns a couple of times. Montgomery (2013) fully explains the statistical analysis of the Latin square design.

A striking real-life example of a 5 × 5 Latin square experiment, designed by Fisher to study the weathering of different species of trees, is given in Plate 6 of Box (1978). This photo may also be viewed online at [19.3]. It shows an aerial view of a huge hillside field in Wales, planted with five different tree species. Plantings are on an altitude gradient across rows within each column and (apparently) on a soil fertility gradient across columns within each row.

Question 19.4

In the field of medicine, a placebo is a pseudo-treatment with no clinically expected therapeutic effect on the patient's specific condition. When clinicians seek to test statistically the effectiveness of a treatment (be it a drug or some other kind of therapy) for a particular condition, a standard aspect of the experimental design is to compare the effect of 'treatment' against 'no treatment'. 'No treatment' can be interpreted literally as overt omission to treat, or treatment can be simulated by administering a placebo.

It is often easier to get patients' cooperation to participate in the experiment by offering all participants some intervention. That is why such a clinical experiment most commonly involves a comparison between the two groups of patients, of treatment versus placebo. Because it has been observed that patients receiving a placebo often feel better, and sometimes even show an actual physiological improvement in their condition (the so-called 'placebo effect', whose mind-body mechanism is still not well understood), the experiment may be done 'single-blind' – that is, a patient is not told whether he or she is receiving the treatment or a placebo. This approach has been shown to lessen the intensity of any placebo effect that might arise. In addition, to avoid possible clinician bias in allocating each patient to the treatment or the placebo group, the experiment may be done 'double-blind'.

Administering a placebo, rather than a clinical treatment, to an unaware patient evidently represents an act of deception by the clinician. This has negative ethical implications, which are all the more serious if the patient's condition is life-threatening. In such circumstances, a code of medical ethics ought certainly to impose a caveat (i.e. a warning or caution) on the use of a placebo.

Both 'placebo' and 'caveat' are Latin verb forms taken directly into English. 'Placebo' (I shall please) is the future indicative of 'placere', to please. This meaning hints at the fact that doctors sometimes prescribe a placebo to keep a patient happy that 'something is being done' when symptoms are minor and self-limiting. 'Caveat' (let him beware) is the present subjunctive of 'cavere', to beware. This reflects its historical origin as a legal warning.

Question 19.5

Some disciplines where experimental studies predominate are chemistry, physics, psychology, pharmacology and agriculture. Some disciplines where observational studies predominate are cosmology, meteorology, climatology, sociology and ornithology. Some disciplines where experimental and observational studies are both common include medicine, biology, geology, economics and education.

What can we conclude? All the sciences have a strong preference for controlled experimentation (for example, comparing the effects of an intervention with the effects of no intervention, while neutralising nuisance variation as far as is possible). Controlled experimentation assists: (i) in clarifying which variables are directly influential in observed relationships; and (ii) in exploring the direction of causality in such observed relationships. Observational studies are inferior for achieving these ends. Fields where observational studies predominate are those where experimentation is either impossible or very difficult.

References

Print

Bennett, J.H. (ed, 1971–1974). *The Collected Papers of R.A. Fisher*. University of Adelaide, 5 volumes.

Box, J.F. (1978). *R. A. Fisher – The Life of a Scientist*. Wiley.

Box, J.F. (1980). R.A. Fisher and the design of experiments, 1922–1926. *The American Statistician* **34**, 1–7.

Fisher, R.A. (1925). *Statistical Methods for Research Workers*. Oliver and Boyd.

Fisher, R.A. (1935). *The Design of Experiments*. Oliver and Boyd.

Lindley, D.V. (1993). An analysis of experimental data – the appreciation of tea and wine. *Teaching Statistics* **15**, 22–25.

Montgomery, D.C. (2013). *Design and Analysis of Experiments*, 8th edition. Wiley.

Senn, S. (2012). Tea for three. *Significance* **9**(6), 30–33.

Online

[19.1] http://www.rothamsted.ac.uk

[19.2] Fisher, R.A. (1933). The contributions of Rothamsted to the development of the science of statistics. In: Bennett, J.H. (ed, 1971–1974). *The Collected Papers of R.A. Fisher*. University of Adelaide. At: https://digital.library.adelaide.edu.au/dspace/bitstream/2440/15213/1/103.pdf

[19.3] www.york.ac.uk/depts/maths/histstat/images/latin_square.gif

Answers – Chapter 20

Question 20.1

Thomas Bayes is buried at Bunhill Fields Burial Grounds, City Road, in the heart of the City of London. Some 200 metres to the west is Errol Street, where the Royal Statistical Society has its office. There are more details online at [20.2].

Question 20.2

In Australian courts, and those of many (but not all) other countries, the accused is presumed innocent until a verdict is announced after a trial. Thus, when the prosecution's evidence is first presented, the finding of AB–blood at the scene is appropriately expressed as $P(E|I) = 0.01$. In the prosecutor's statement this is incorrectly switched to mean $P(I|E) = 0.01$, and therefore $P(G|E) = 0.99$. That certainly enhances the prosecution's case! However, it is a logical mistake, quite reasonably called the 'prosecutor's fallacy'. Unless the mistake is picked up by others involved in the trial, it may lead to a miscarriage of justice. Examples of the prosecutor's fallacy in some famous trials are given in the Wikipedia article at [20.3].

Question 20.3

Let A_i be the event that there are initially i red balls in the bin. The answer to this question depends on what assumption we make about the value of the prior probabilities $P(A_i)$ for $i = 0, 1 ..., 10$.

Since we have no prior information on how the bin was initially filled, a simple option is to follow the 'principle of insufficient reason' (alternatively known as the 'principle of indifference'), and assume that all the events, $A_0, A_1, ..., A_{10}$, are equally likely. Then $P(A_i) = 1/11$. Next, using r to denote the selection of a red ball, we apply Bayes' theorem: $P(A_1|r) = P(r|A_1).P(A_1)/\Sigma[P(r|A_i).P(A_i)]$. Substituting assumed values:

$$P(A_1 | r) = [0.1 \times 1/11] / \Sigma[(i/10) \times (1/11)] = 1/55.$$

Another possibility for defining the values $P(A_i)$ is to assume that the bin was initially filled with a random selection of 10 balls from some vast reservoir containing red and black balls in equal numbers. This implies a binomial distribution for A_i, with $P(A_i) = {}^{10}C_i (0.5)^i (0.5)^{10-i} = {}^{10}C_i (0.5)^{10}$. Again applying Bayes' Theorem:

$$P(A_1 | r) = [0.1 \times {}^{10}C_1 (0.5)^{10}] / \Sigma[(i/10) \times {}^{10}C_i (0.5)^{10}] = 1/512.$$

(See the note on incompletely formulated probability problems in the answer to QUESTION 10.5(b).)

Question 20.4

As mentioned in the Overview, the brilliant British logician Alan Turing (1912–1954), with other gifted colleagues, broke several increasingly elaborate coding mechanisms of the Enigma text-enciphering machine, used to convey orders to the field by the German Military High Command through-out World War II. This work was done in top secret offices located in a mansion, known as Bletchley Park, in the English county of Buckinghamshire. Its successes remained secret, moreover, until the 1970s. The task was accomplished by an extended process of trial-and-error hypothesising. At each stage, the strength of evidence favouring some particular coding mechanism was revised (in Bayesian fashion) in the light of accumulating evidence (e.g. from newly intercepted Enigma messages).

Turing's invented scale for measuring strength of evidence was the loga-rithm (base 10) of the odds ratio in favour of a particular hypothesis about the coding mechanism. The unit on this scale corresponds to odds of 10 to 1. This, Turing called '1 ban'. Work on this project often involved rather lower favourable odds – for example, 5 to 4, equivalent to 0.10 ban, or 1 deciban. When the odds of a particular hypothesis strengthened to 50 to 1, or 1.70 ban, the analysts decided that they were close enough to be sure that they were correct.

Jack Good, Turing's statistical assistant at Bletchley Park, reveals in his memoir (1979, p. 394) the marvellous information that 'a deciban or half-deciban is about the smallest change in weight of evidence that is directly perceptible to human intuition.'

The name 'ban' derived from the nearby town of Banbury, where a printing shop supplied large quantities of stationery to the Enigma decod-ing project. McGrayne describes the whole project in detail in chapter 4 of her book. For a short overview of events at Bletchley Park, see Simpson (2010).

Question 20.5

In the frequentist theory of inference, the population parameter to be esti-mated (e.g. the mean) is treated as a *fixed* value. The sample mean is treated as a *random* variable. Values of the sample mean are generated by repeated sampling from the population. These values compose the sampling distri-bution of the sample mean. Then, a 95% confidence interval (CI) for the population mean is determined by finding a range of values that embraces 95% of sample means. (Note: a 95% CI constructed in this way is not unique. A further restriction is required to make it so – e.g. equal tails.) The fre-quentist approach to interval estimation is characteristically tied to the notion of *a probability distribution of sample means*.

Since the population mean is regarded as a fixed value, *it is not valid* to interpret a 95% CI (e.g. 2.5 to 3.5) by saying there is a 95% probability that the population mean lies between 2.5 and 3.5. Either it does or it doesn't.

The probability is one or zero. Instead, the 95% CI may be interpreted as follows. We do not know whether the population mean lies within the calculated interval but we may act, in practice, as if it does, because the interval estimator (formula) is successful in capturing the mean in 95% of samples from the population to which it is applied.

In the Bayesian theory of inference, the population parameter to be estimated (e.g. the mean) is treated as a *random* variable. At the outset, there may be extrinsic factual evidence or personal hunches (i.e. subjective ideas) as to the value of the population mean. All such diffuse information can be summarised in a prior distribution of values of the population mean (if no specific prior information is available, the uniform distribution will serve as the prior distribution). Next, a *fixed* (i.e. single) sample is obtained from the population. Given the value of the sample mean, the prior distribution of population mean values can be revised by applying Bayes' theorem to the prior probabilities. The revision is termed the posterior distribution of population mean values. Finally, a 95% probability interval for the population mean is defined directly on the posterior distribution. Such a Bayesian probability interval is commonly termed a 'credible interval' (note: as before, this interval is not unique). The Bayesian approach to interval estimation is characteristically tied to the notion of *a probability distribution of likely values of the population mean.*

Because the 95% credible interval is a probability interval obtained directly on the posterior distribution of population means, *it is valid* to say that the probability is 95% that the population mean lies within the credible interval. True, the Bayesian interval may be more complex to construct than the frequentist interval, but it has a simpler and more intuitively appealing interpretation.

For a broader and reasonably accessible account of Bayesian inference, see O'Hagan (2008), online at [20.4]. For evidence of growing harmony in the views on inference of frequentists and Bayesians, see Kass (2009), online at [20.5].

References

Print

Good, I.J. (1979). Studies in the history of probability and statistics XXXVII.
A.M. Turing's statistical work in World War II. *Biometrika* **66**, 393–396.
McGrayne, S. (2012). *The Theory That Would Not Die.* Yale University Press.
Simpson, E. (2010). Bayes at Bletchley Park. *Significance* 7(2), 76–80.

Online

[20.2] http://bayesian.org/bayes

[20.3] https://en.wikipedia.org/wiki/Prosecutor's_fallacy

[20.4] O'Hagan, A. (2008). The Bayesian Approach to Statistics. Chapter 6 in: Rudas, T. (ed, 2008). *Handbook of Probability: Theory and Applications.* Sage. At: http://www.sagepub.com/upm-data/18550_Chapter6.pdf

[20.5] Kass, R.E. (2009). Statistical inference: the big picture. *Statistical Science* **26**, 1–9 (with further discussion by others on pages 10–20). At: https://projecteuclid.org/euclid.ss/1307626554

Answers – Chapter 21

Question 21.1

The phrase is ETAOIN SHRDLU, the 12 most frequent letters in English prose, in order. SHRDLU was used (sometimes repeatedly) by linotype operators to close off a line of type as soon as they saw they had made a typographical error. This acted as a signal to proof readers to remove that line physically from the page form. (English language linotype machines – used until the 1970s for 'hot metal' typesetting of newspapers and other printed publications – had a keyboard layout differing from the conventional QWERTY keyboard, in that adjoining keys bore letters in their frequency order of occurrence. SHRDLU could be set by a deft downwards glissando of a finger over six keys, all in a single column!)

The phrase, or parts of it, have appeared in many other contexts for more than a century. Shrdlu was the name of an early artificial intelligence system developed, as part of his doctoral studies, by Terry Winograd at MIT in the late 1960s (see [21.2]) and, a decade later, Douglas Hofstadter (1979) included in his book *Gödel, Escher, Bach* a dialogue between a computer programmer Eta Oin and the Shrdlu program. Other uses are mentioned by Michael Quinion on his World Wide Words site at [21.3], and there is also an extensive list in the Wikipedia entry at [21.4].

Question 21.2

The map was created by Charles Booth (1840–1916), an English businessman and social reformer. Booth had a deep Victorian sense of obligation towards the poor and the improvement of their living conditions. In his youth, he was dissatisfied with the quality of contemporary data on social affairs so, from 1886 to 1903, he organised his own extensive survey, and published the results in several volumes under the title *Inquiry into the Life and Labour of the People in London*. The data were collected by a small team of investigators who visited every part of London, interviewing people in their homes or workplaces. Booth was the first to use the phrase 'line of poverty' (now called the 'poverty line'), and concluded that as many as 35% of the population were at or below this line.

The *Maps Descriptive of London Poverty* are an early example of social cartography, with each street coloured to show the socio-economic status of the inhabitants. Much more information (and an alternative version of the map) is available at the Charles Booth Online Archive at [21.5]. Volume 1 of Booth's report can be read online at [21.6].

We took the map from David Thomas' website, 'Charles Booth's 1889 descriptive map of London poverty' at [21.7], and the partial view reproduced in FIGURE 21.1 is from [21.8]. It depicts the area north of Piccadilly, and was chosen whimsically for its inclusion of Broad Street, a street made famous by events referred to in QUESTION 1.5.

The current statistical marketing technique of geodemographic segmentation – dividing a market into groups, according to geographic location and demographic and economic variables (such as level of education and family income) – seems to be a direct descendant of Booth's work. See, as an example, a technical briefing paper by Abbas and colleagues at [21.9].

Question 21.3

a) The chi-squared test of independence does not use any information about the order in the responses, which is equivalent to treating the response as a categorical variable. The t-test uses the numerical values of the responses, which is equivalent to treating the response as a quantitative variable. Neither of these is appropriate, strictly speaking, since the response is actually an ordinal variable. The non-parametric Mann-Whitney test is preferable for comparing two independent groups on an ordinal variable.

b) All of the tests in (a) would throw some light on whether the two treatments were equally effective, assuming that the patients' responses represented the actual level of effectiveness (as opposed to perceived effectiveness). If the null hypothesis of the chi-squared test were rejected, we would have evidence that the *pattern* of effectiveness was different in the two groups. If the null hypothesis of the t-test or the Mann-Whitney test were rejected, we would have evidence of different *average* effectiveness in the two groups. It is important to recognise that a failure to reject the null hypothesis in any of the three tests could be due to a lack of statistical power to distinguish any actual difference in effectiveness, and should not be immediately interpreted as evidence that the treatments are equally effective.

Question 21.4

The paper is by Campbell and Joiner (1973). The type of sampling described in this paper is called 'randomised response'. This technique is a way of getting information on 'difficult' questions (such as recreational drug taking or sexual habits) without having respondents incriminate or embarrass themselves. Two questions are posed by the interviewer – the difficult one and a harmless one – and a reply to one of these questions is requested. The actual question which the respondent answers is chosen by the outcome of a chance experiment (such as the toss of a coin) that is performed by the respondent but is not revealed to the interviewer. More information on the randomised response technique can be found in Fox and Tracy (1986).

Question 21.5

During World War II, the statistician Abraham Wald (1902–1950) was employed within a US Government research group based at Columbia University. In 1943, Wald undertook a statistical modelling project designed to improve the survivability of military aircraft in combat, by determining which specific parts of an aircraft's structure would most benefit from increased armour-plating. The analysis was greatly complicated by two fundamental factors:

i) little was known in reality about the pattern of vulnerability of different regions of an aircraft's structure to assured destruction from a single bullet strike, and
ii) no data were available from those aircraft that did not return from their missions. A further obstacle was the difficulty, in that pre-computer era, of performing the nonlinear constrained optimisation calculations to which Wald's analysis led him.

In the course of this work, Wald determined approximate confidence intervals for the probabilities that an aircraft would survive a hit on each of several of its structural parts. Inspection of the returned aircraft revealed that the parts of their structure corresponding to the lowest probability of aircraft survival were, in fact, significantly less damaged than other parts of those aircraft. Wald surmised that the aircraft that did not return were lost precisely because they were hit in those *theoretically* vulnerable places that were found *in practice* to be mostly undamaged in the aircraft that returned. To the initial consternation of many, Wald recommended that the surviving aircraft be reinforced in those places that had gone mostly undamaged in previous missions.

References

Print

Campbell, C. and Joiner, B. (1973). How to get the answer without being sure you've asked the question. *The American Statistician* **27**, 229–231.

Fox, J.A. and Tracy, P.E. (1986). *Randomized Response: A Method For Sensitive Surveys*. Sage.

Hofstadter, D. (1979). *Gödel, Escher, Bach: An Eternal Golden Braid*. The Harvester Press.

Online

[21.2] http://hci.stanford.edu/winograd/shrdlu/

[21.3] http://www.worldwidewords.org/weirdwords/ww-eta1.htm

[21.4] http://en.wikipedia.org/wiki/ETAOIN_SHRDLU

[21.5] http://booth.lse.ac.uk/

[21.6] http://www.archive.org/details/labourlifeofpeop01bootuoft

[21.7] http://www.umich.edu/~risotto/

[21.8] http://www.umich.edu/~risotto/maxzooms/nw/nwe56.html

[21.9] Abbas, J. *et al.* (2009). *Geodemographic Segmentation*. Association of Public Health Observatories. At: www.apho.org.uk/resource/view.aspx?RID=67914

Answers – Chapter 22

Question 22.1

a) The statement (its exact words are italicised below) first appeared in a passage ostensibly extracted from a longer (but, in fact, non-existent) literary work titled 'The Undoing of Lamia Gurdleneck' and authored by K.A.C. Manderville. Here is the text:

> 'You haven't told me yet,' said Lady Nuttal, 'what it is your fiancé does for a living.' 'He's a statistician,' replied Lamia, with an annoying sense of being on the defensive. Lady Nuttal was obviously taken aback. It had not

occurred to her that statisticians entered into normal social relationships. The species, she would have surmised, was perpetuated in some collateral manner, like mules. 'But Aunt Sara, it's a very interesting profession,' said Lamia warmly. 'I don't doubt it,' said her aunt, who obviously doubted it very much. 'To express anything important in mere figures is so plainly impossible that there must be endless scope for well-paid advice on how to do it. But don't you think that life with a statistician would be rather, shall we say, humdrum?' Lamia was silent. She felt reluctant to discuss the surprising depth of emotional possibility which she had discovered below Edward's numerical veneer. *'It's not the figures themselves,'* she said finally, *'it's what you do with them that matters.'*

This passage is a sharp-eyed parody of 19th century fiction, as well as an elaborate joke. It was concocted by Maurice Kendall and Alan Stuart, authors of *The Advanced Theory of Statistics* (in three volumes), to serve as a whimsical epigraph to volume 2 of that treatise. Multiple anagrams are present here. 'Lamia Gurdleneck' is revealed as 'Maurice G. Kendall' in another guise, just as 'K.A.C. Manderville' is an alter-ego of 'Mavrice Kendall.' Similarly, 'Sara Nuttal' turns out to be none other than 'Alan Stuart' in disguise.

b) The US statistician W.J. Youden (1900–1971) spent the first half of his scientific career as a plant chemist, and the second half as an applied statistician at the US National Bureau of Standards. He created a lyrical typographic tribute to the normal distribution. In FIGURE 26.8, we replicate his wording, if not also the perfection of his typesetting.

THE
NORMAL
LAW OF ERROR
STANDS OUT IN THE
EXPERIENCE OF MANKIND
AS ONE OF THE BROADEST
GENERALIZATIONS OF NATURAL
PHILOSOPHY • IT SERVES AS THE
GUIDING INSTRUMENT IN RESEARCHES
IN THE PHYSICAL AND SOCIAL SCIENCES AND
IN MEDICINE AGRICULTURE AND ENGINEERING •
IT IS AN INDISPENSABLE TOOL FOR THE ANALYSIS AND THE
INTERPRETATION OF BASIC DATA OBTAINED BY OBSERVATION AND EXPERIMENT

Figure 26.8 Youden's tribute to the normal.

The original appears on page 55 of his nontechnical book for students, Youden (1962). Scanned reproductions of this book can be found at several sites on the web.

Question 22.2

The original diagram was constructed in 1858, following the Crimean War of 1854–56, by Florence Nightingale (1820–1910), nurse, administrator, hospital reformer, public health advocate and applied statistician, to persuade the British government that improvements in hospital hygiene could save lives (in particular, the lives of soldiers). The diagram contrasts, month by month, the number of soldiers who died in battle with the much greater number who died subsequently in military hospitals of preventable causes (chiefly, sepsis of their wounds). In this diagram, invented by Nightingale and now technically termed a polar-area diagram, magnitudes are represented by areas. The result is a more dramatic display of the data than a bar chart presents.

In some accounts of Nightingale's contributions to effective statistical graphics, the diagram has been called a 'coxcomb', apparently because this term occurs in her own correspondence. However, her biographer, Hugh Small, has shown that this is a misattribution. In a paper for a 1998 conference organised by the Florence Nightingale Museum in London, Small refers to the diagram as a 'wedge diagram'. The paper is online at [22.11]. We constructed FIGURE 22.3 using the *R* programming language and the information in the notes on the *HistData* package, maintained by Michael Friendly [online at 22.12].

A captivating account of Nightingale's statistical activities, structured as an imagined dialogue with her, is in Maindonald and Richardson (2004), online at [22.13].

It is worth noting that Nightingale was the first female Fellow of the Royal Statistical Society.

Question 22.3

a) The 'trimmed mean' is obtained by ordering the data by size, discarding some extreme values at each end, and then calculating the arithmetic mean of the remaining values. For example, to find the 10% trimmed mean, one discards the largest 5% and the smallest 5% of the data values, and calculates the arithmetic mean of those that remain. Clearly, a trimmed mean is less sensitive (i.e. more robust) than the arithmetic mean to outliers in the data. That can be an advantage when the data are suspected to contain large observational or measurement errors.

b) In 1972, Tukey, who had already had an interest in robust estimation for more than a decade, published (with five co-authors) a book-length study, titled *Robust Estimates of Location* (see Andrews *et al.*, 1972). Because this group of researchers were all, at that time, affiliated with

Princeton University, the study became known as the Princeton Robustness Study. The objective of the study was to examine more than 60 alternative point estimators of the central value of a symmetrically distributed population in order to determine how robust they are under various departures from the 'well-behaved' estimation contexts of statistics textbooks. These departures included populations defined by non-normal standard distributions (including some asymmetric distributions) and mixtures of standard distributions; and different patterns of outliers among the sample data.

About a third of the estimators reviewed were trimmed means or elaborate variants of trimmed means. Other statistics evaluated included maximum likelihood-type estimators for populations of known form, the median, and elaborate variants of the median. The estimators were compared on several criteria, including: asymptotic bias and efficiency; bias and efficiency in small samples; and ease of computation. In a summary (page 254) of the mass of results in the book, the 10%, 15% and 25% trimmed means and several maximum likelihood-type estimators were found to have performed very well in the widest array of contexts. In sharp contrast, the sample mean, which is optimal on so many criteria in textbook estimation of the mean of a normal population, was actually the worst performer of all in 'messy' data!

Question 22.4

The statistician was Francis Galton and the fair at which he came upon an ox as the subject of a weight-guessing competition took place in 1906. Galton (1907), online at [22.14], reported what he discovered when he analysed the guesses that were submitted, in an article titled *Vox populi*. From the 787 weight guesses that he had, he found the median to be 1207 pounds. Galton reasoned that the median was, in this context, the best estimate one could use of the unknown dressed weight of the ox, for there would, in principle, be a majority opinion against any other estimate as being either too high or too low. The ox's dressed weight was actually 1198 pounds. Thus, the median was in error by less than 1%. This finding was, he concluded, 'more creditable to the trustworthiness of democratic judgment than might have been expected'.

In the past ten years, Galton's simple investigation, prompted by nothing more than casual curiosity, has been elaborately reinterpreted. That an average of many people's quantitative evaluations in *any* sampling situation (random or not) should prove more accurate than that of most of those people individually has been dubbed 'the wisdom of crowds'. An extensive literature, both scholarly and popular, has emerged, seeking to put this supposedly substantial phenomenon on a firm theoretical foundation in the fields of psychology, economics and even genetics. This literature includes enormously controversial claims and similarly robust rebuttals. It is not yet clear what the net judgment will be on the wisdom of crowds.

Question 22.5

a) Gosset began work as a chemist at the Guinness Brewery in Dublin in 1899, and spent his entire career with that company. The formal title of his position was 'Brewer', and it was a job of great status for Guinness was initially the only Irish brewery to engage trained scientists to oversee production. Between September 1906 and June 1907, he studied statistics in London with Karl Pearson, then a professor at University College. In taking this time out for academic study, Gosset had a particular goal (among several others). It was to be able to test hypotheses about a mean based on a small sample of data, which was all that he had in his work on improving methods of beer brewing. Since no one had yet explored this topic in statistical inference (because everyone relied on the use of large sample normal approximations), Gosset set out to do the pioneering work himself.

    When he first contemplated publishing the fruits of his research in scholarly journals, he sought the consent of his employer. The Guinness Board agreed, provided that he adopted a pseudonym and included no data identifiable as from Guinness. This demand for secrecy was dictated by the fact that his work was regarded as (what we today term) 'commercial in confidence'. This Board ruling applied to all Guinness' scientifically-skilled employees. Gosset adopted the pseudonym 'Student' when he published his first paper in 1907 (we may now guess) precisely because he saw himself at that time as a student of statistics. There is much vivid detail about Gosset in this era in Box (1987), online at [22.15].

b) It should be clear that this question is not about which letter, z or t, was used in the work done by Student and by Fisher. It is a question about the process of eponymy in statistics – how some concepts, constructs or techniques become named after people. Eponymy is evidently intended as a way of honouring the contributor, but the process has no formal rules, and little consistency from case to case. You might, at first sight, think that the originator of some influential advance in theory or practice would become the eponym. But what if: (i) it subsequently turns out that some other person had earlier made the same advance, which was then lost to view and has been independently rediscovered; or (ii) the originator had a remarkable insight or intuition that later proved to be very fertile, though he or she did not realise it, or did not succeed in demonstrating it? In such cases, who should be the eponym? There are, indeed, many contexts in statistics where the eponym is not at all the person you might think it *should* be, but few contexts in which collective disapproval actually resulted in an alteration of the eponym.

On the centenary of Student's 1908 paper, Zabell (2008) reviewed these issues (and others) in the eventful technical and interpersonal history of the *t*-distribution. The 'Discussion' published in the pages immediately following Zabell's article is also particularly informative. It turns out that both the circumstances mentioned in the previous paragraph apply.

In his work on inference from small samples, Student had two forerunners – the German statistician Jacob Lüroth (1844–1910) in 1876 and Francis Ysidro Edgeworth, mentioned in Figure 22.1, in 1883. And Student (in working out the distribution of his *z*-statistic) did not provide a complete formal proof. Fisher supplied that in 1923. Nor did Student foresee the possibility of unifying a whole class of hypothesis tests, as Fisher demonstrated with his *t*-distribution. So, though it has been Student's distribution in every textbook now for at least 75 years, *should* it be Fisher's ... or Lüroth's? Once again, we invite you to be the judge.

On the meaning of 'probable error', as it appears in the title of Student's 1908 paper, see the historical note following the answer to Question 6.1. There is more in Chapter 23 about eponymy in statistics.

References

Print

Andrews, D. F., Bickel, P. J., Hampel, F. R., Huber, P. J., Rogers, W. H., and Tukey, J. W. (1972). *Robust Estimates of Location: Survey and Advances*. Princeton University Press.

Youden, W.J. (1962). *Experimentation and Measurement*. [US] National Science Teachers' Association.

Zabell, S.L. (2008). On Student's 1908 article 'The probable error of a mean'. *Journal of the American Statistical Association* **103**, 1–7. With Discussion, 7–20.

Online

[22.11]   http://www.york.ac.uk/depts/maths/histstat/small.htm

[22.12]   https://cran.r-project.org/web/packages/HistData/HistData.pdf

[22.13] Maindonald, J. and Richardson, A. (2004). This passionate study: a dialogue with Florence Nightingale. *Journal of Statistics Education* **12**(1). At: www.amstat.org/publications/jse/v12n1/maindonald.html

[22.14] Galton, F. (1907). Vox populi. *Nature* **75**, 450–451. At: http://galton. org/essays/1900-1911/galton-1907-vox-populi.pdf

[22.15] Box, J.F. (1987). Guinness, Gosset, Fisher and small samples. *Statistical Science* **2**, 45–52. At: https://projecteuclid.org/download/ pdf_1/euclid.ss/1177013437

Answers – Chapter 23

Question 23.1

i) The Behrens-Fisher test is a parametric test of the difference of means of two normal populations with unknown *but unequal* variances, based on data from two independent random samples. Devising such a test is a more intractable problem than in the case where the unknown variances are equal. In 1929, a German agricultural scientist, Walter Behrens (1902–1962), explained why it was difficult to construct an exact test in the unequal variances case and suggested a possible solution. Fisher (1939), online at [23.3], citing Behrens, offered a different solution. The joint eponymy became established in the 1940s. See 'Behrens-Fisher' on Jeff Miller's website on the early history of terms in mathematics and statistics, at [23.4].

ii) The Durbin-Watson test is a small-sample bounds test for first-order autoregression in the disturbances of a regression model. The test statistic is calculated from the regression residuals after ordinary least squares estimation. The theory underlying the test was published in two articles in *Biometrika* in 1950 and 1951. The joint eponymy originated shortly after these articles appeared. The English statistician James Durbin (1923–2012) collaborated in developing the test with the Australian-born statistician Geoffrey Watson (1921–1998), who worked for many years at Princeton University. Durbin describes the background to their work in Phillips (1988). An exact small-sample test of the same hypothesis is now available.

iii) The Wilcoxon signed-ranks test, introduced in Wilcoxon (1945), is a non-parametric test of the difference of means of two populations, using matched samples. Thus, it is a non-parametric alternative to Student's paired *t*-test. Frank Wilcoxon (1892–1965) was a chemist and statistician in US industry and academia.

Question 23.2

i) The Russian statistician Pafnuti Lvovich Chebyshev (1821–1894) was not the originator of Chebyshev's inequality. Stigler, on page 283 of his book *Statistics on the Table*, points out that Chebyshev was anticipated by the French statistician Irénée-Jules Bienaymé (1796–1878), who established the same inequality some 15 years earlier, and in greater generality than Chebyshev. For further detail on Bienaymé's priority in this area, see Heyde and Seneta (1977), pages 121–124.

ii) In the late 19th century, the usual way of analysing the income distribution within a country was by fitting an appropriate statistical model to the empirical frequency distribution of numbers of income earners by level of income. This approach, however, is inadequate to show income *relativities* – that is, the degree to which total personal income is concentrated among those at the top of the income scale, relative to those lower down. Income relativities became important where economic policy-makers wanted to restrain the growth of income inequality across the nation.

One can show income relativities graphically by plotting cumulative income against cumulative numbers of income earners. To permit inter-country comparisons, where currencies differ, the cumulations need to be in percentage terms (so freeing them of the unit of measurement). The end result is the curve of cumulative percentage of total income, plotted against cumulative percentage of income earners (arranged in order from the lowest earners to the highest).

Inventing this plot (it is found in Lorenz, 1905) was the valuable statistical contribution of Max Lorenz (1876–1959), a US applied economist and statistician. It became known quite quickly as the Lorenz curve (see, for example, Watkins (1909), page 172), and continues today to be a standard tool in the study of national distributions of both income and wealth.

Was Lorenz the originator of the Lorenz curve? He seems to have been the first to announce it in print. Others (apparently independently) presented similar diagrams soon after. Derobert and Thieriot (2003) and Kleiber (2007) cite work by Chatelain in France in 1907, and by Pietra in Italy in 1915. So this is likely to have been an instance of what Merton (1973), chapter 16, calls 'multiple discovery'.

iii) Stigler (1983) casts doubt on the proposition that Thomas Bayes (1702–1761) is the originator of Bayes' theorem, but Stigler's ingeniously-elicited evidence in favour of the English mathematician Nicholas Saunderson (1682–1739) is suggestive, rather than conclusive.

Question 23.3

Von Bortkiewicz's data, collected from official sources in 1898, relate to the number of deaths, per individual army corps, of Prussian Army soldiers from horse kicks. His data span the 20 years 1875–1894 for each of 14 cavalry corps. The full data set of 280 values is given in Hand *et al.* (1994). Von Bortkiewicz argued heuristically that the Poisson probability model was a good fit to 200 of these data values (excluding four heterogeneous corps). It was not until 1925 that R.A. Fisher demonstrated the good fit via the chi-squared test.

Poisson so underestimated the broad practical value of the probability distribution that bears his name, says Good (1986, page 166), online at [23.5], that it ought to have been named the von Bortkiewicz distribution.

When did the Poisson distribution get its name? The entry for 'Poisson distribution' (online at [23.6]) on Jeff Miller's website, already referred to in the answer to QUESTION 23.1, indicates that this term was first used in a 1922 journal article, of which R.A. Fisher was the primary author. Thus, Poisson, who first presented his distribution in 1830 (see Dale, 1988), had to wait almost a century to acquire his status as an eponym.

See QUESTION 13.3 for more on this remarkably versatile probability model.

Question 23.4

Chernoff's picture is a cartoon of a human face, with the possibility of varying different facial features so as to represent in two dimensions a vector of measurements in up to 18 dimensions (see Chernoff, 1973).

There are now many websites where 'Chernoff faces' are described, illustrated and critiqued. Several software packages have add-ons for producing these faces, including *Mathematica* – see online at [23.7]. Other ways of plotting multivariate data as familiar images have been surveyed by Everitt and Nicholls (1975).

Question 23.5

The paper is Box and Cox (1964), titled 'An analysis of transformations'. It is the work of the British statisticians George Box and David Cox. The novel technique introduced by these authors is now universally known as the Box-Cox transformation. The operetta is Cox and Box, first performed publicly in 1867. Its main characters are James John Cox and John James Box. You can enjoy the music on YouTube. How this operetta led to the statistical collaboration is recounted on pages 254–255 of DeGroot (1987), online at [23.8]. The allusion evoked by their coupled names was, it seems, compelling, as it was the initial spark for the Box-Cox collaboration. The paper's content was decided thereafter!

Incidentally, this case of authors joining together opportunistically for whimsical effect has a precedent – but there, the paper's content was completed first. In 1948, at George Washington University in the USA, a doctoral student in physics, Ralph Alpher, was about to submit for publication a research paper written with his thesis adviser, George Gamow. Gamow, clearly an aficionado of the Greek alphabet, could not resist co-opting, as a third author, the physicist Hans Bethe (apparently overriding Alpher's objection). This led to the publication of Alpher, Bethe and Gamow (1948), subsequently a highly cited paper in cosmology. The 'Alpher-Bethe-Gamow paper' now has its own entry in Wikipedia.

References

Print

Alpher, R.A., Bethe, H. and Gamow, G. (1948). The origin of chemical elements. *Physical Review* **73**, 803–804.

Box, G. and Cox, D. (1964). An analysis of transformations. *Journal of the Royal Statistical Society, Series B* **26**, 211–243.

Chernoff, H. (1973). The use of faces to represent points in k-dimensional space graphically. *Journal of the American Statistical Association* **68**, 361–368.

Dale, A.I. (1988). An early occurrence of the Poisson distribution. *Statistics & Probability Letters* **7**, 21–22.

Derobert, L. and Thieriot, G. (2003). The Lorenz curve as an archetype: A historico-epistemological study. *European Journal of the History of Economic Thought* **10**, 573–585.

Everitt, B.S. and Nicholls, P.G. (1975). Visual techniques for multivariate data. *The Statistician* **24**, 37–49.

Hand, D. *et al.* (1994). *Handbook of Small Data Sets.* Chapman and Hall.

Heyde, C. and Seneta, E. (1977). *I.J. Bienaymé – Statistical Theory Anticipated.* Springer.

Kleiber, C. (2008), The Lorenz curve in economics and econometrics. In: Betti, G. and Lemmi, A. (eds). *Advances on Income Inequality and Concentration Measures: Collected Papers in Memory of Corrado Gini and Max O. Lorenz,* pp. 225–242. Routledge.

Lorenz, M.O. (1905). Methods of measuring the concentration of wealth. *Quarterly Publications of the American Statistical Association* **9**, June, 209–219.

Merton, R.K. (1973). *The Sociology of Science.* The University of Chicago Press.

Phillips, P.C.B. (1988). The ET Interview: Professor James Durbin. *Econometric Theory* **4**, 125–157.

Stigler, S.M. (1983). Who discovered Bayes's Theorem? *The American Statistician* **37**, 290–296. Reprinted with updates as chapter 15 of Stigler (1999).

Stigler, S.M. (1999). *Statistics on the Table: The History of Statistical Concepts and Methods.* Harvard University Press.

Watkins, G.P. (1909). The measurement of concentration of wealth. *Quarterly Journal of Economics* **24**, 160–179.

Wilcoxon, F. (1945). Individual comparisons by ranking methods. *Biometrics* **1**(6), 80–83.

Online

[23.3] Fisher, R.A. (1939). The comparison of samples with possibly unequal variances. *Annals of Eugenics* **9**, 174-180. At: https://digital.library. adelaide.edu.au/dspace/bitstream/2440/15235/1/162.pdf

[23.4] http://jeff560.tripod.com/b.html

[23.5] Good, I.J. (1986). Some statistical applications of Poisson's work. *Statistical Science* **1**, 157–180. At: https://projecteuclid.org/euclid. ss/1177013690

[23.6] http://jeff560.tripod.com/p.html

[23.7] http://mathworld.wolfram.com/ChernoffFace.html

[23.8] DeGroot, M.H. (1987). A conversation with George Box. *Statistical Science* **2**, 239–258. At: https://projecteuclid.org/euclid.ss/1177013223

Answers – Chapter 24

Question 24.1

As we mention in CHAPTER 14, Gauss and Laplace both advocated the normal as a probability model for the distribution of random errors of measurement, because its main shape characteristics matched those of corresponding empirical frequency distributions. Many statisticians confirmed that the normal was an excellent model for such data. On that evidence, they dubbed it the normal 'law of error'.

However, we should not overlook that Gauss also favoured the normal distribution because it has desirable properties for statistical theory that are mathematically provable. As stated in this chapter's Overview, it is the only symmetric continuous pdf for which the mean of a random sample is the maximum likelihood estimator of the population mean. That made choosing the normal as the 'law of error' doubly appealing to Gauss.

Thus, the normal law of error was not solely deduced from some abstract mathematical theorem; nor was it solely induced from some experimentally determined facts. Gauss endorsed it as a happy fusion of both.

Incidentally, you can read the original version of Poincaré's anecdote on the online page at [24.2].

Question 24.2

The fact that there is a markedly unequal distribution of leading digits in many large real-life collections of numbers seems quite counterintuitive – all the more so, when other collections (e.g. lottery results and national postcodes) do not display the same behaviour. The mathematician Simon Newcomb was, in 1881, the first to write about this phenomenon, which he inferred from the more worn early pages of books of logarithm tables. There the topic rested, until it was explored at length by the physicist Frank Benford in 1938. Thereafter, various approaches to proving a probability law for the observed unequal relative frequencies of leading digits were tried.

Both Newcomb and Benford conjectured the following result (now known as Benford's Law): the probability that the first digit in a very large collection of numbers from diverse sources is n is $\log_{10}[(n + 1)/n]$. Thus, the probability of 1 is 0.30, of 2 is 0.18, and so on. It is only relatively recently that a formal proof has confirmed the validity of this surprising result – see the (advanced) paper by Hill (1995), online at [24.3]. A non-technical account of some ideas behind this proof is in Hill (1998), online at [24.4], and a popular overview of Benford's Law is in Walthoe (1999), online at [24.5].

How can you know in advance whether a particular collection of numbers will or will not follow Benford's Law? This is still an open question, to which Fewster (2009) offers some constructive answers.

Incidentally, the naming of this law for Benford, after it was first announced by Newcomb 57 years earlier, is another confirmation of Stigler's Law of Eponymy, which you will find in CHAPTER 23.

Ever since financial data were found to align well with Benford's Law, it has found a fertile field of application in forensic accounting. Commercial fraud may involve the creation of false business data, which perpetrators hope to pass off as genuine. However, such data are likely to have leading digits in proportions that diverge from those predicted by Benford's Law. Hence, statistical tests on digit frequency may point auditors to anomalous data worth deeper investigation. A proponent of using statistical analysis, and Benford's Law in particular, to detect fudged accounts is the US academic Mark Nigrini. He popularised his approach in Nigrini (1999) and subsequently published a reference book for professionals, Nigrini (2012).

The practical value of Benford's Law for detecting numerical fudging in other scientific fields is currently being explored.

Question 24.3

The remarkable versatility of the normal distribution as a statistical model for observed frequencies of many *continuous* real-world variables is well known. It serves well for roughly symmetrical raw data in the physical and biological sciences, and – through the CLT – for the large-sample distribution of sample means, even from non-symmetric populations. However, real-world variables – both continuous and discrete – that are clearly not (even approximately) normal are also common, especially in the social and behavioural sciences. As this chapter's Overview shows, some such continuous variables have a two-tailed distribution, with tails that are fatter than the normal's. These are often modelled by one of the family of stable distributions.

Then there are other – typically, discrete – variables that have a distribution comprising many small values but few very large values, and (in contrast to the normal) few values that fall near the mean. Such a distribution is one-tailed with a roughly L-shaped profile.

Since the 19th century, it has been known that a good model for the frequency, $f(x)$, of a *discrete* random variable, x, with such a distributional profile, is likely to be found in the general form $f(x) = ab^x$, where $x \geq 0$, a is a positive constant of proportionality (to convert $f(x)$ into a pdf – i.e. to make the probabilities sum to 1) and b is a parameter in the range $0 < b < 1$. This model is known as the geometric distribution. It follows from this definition that the geometric distribution is likely to be a good model for a set of empirical data if the plot of $\log(f(x))$ against x resembles a straight line. Other long-established models with a related general form are the Poisson distribution (for discrete variables) and the exponential distribution (for continuous variables).

However, while data often fit well to a geometric distribution near the mode of the empirical distribution, the fit in the tail is sometimes poor. In the decades before 1960, several researchers, each in a separate field, discovered independently that a better overall fit than the geometric distribution provided could be had by using an alternative model. This alternative also generated a broadly L-shaped profile, but one with a subtly different curvature. In some applications, this model was fitted to a raw frequency distribution, and in others to a cumulative frequency distribution. Again, sometimes it was used to model a discrete variable, and sometimes a continuous variable. In each field, the model received a different name. Some examples are Lotka's Law, Pareto's Law and Bradford's Law.

By 1960 it was realised that all these individual probability laws were simply variants of a common form. It was time for a unified terminology! Today, the common form is known as the *power law* (or, more correctly, the family of power laws).

If a *discrete or continuous* random variable, *x*, has a distribution that follows a power law, then its frequency or density function, respectively, is given by $f(x) = ax^{-b}$, where $x > 0$, *a* is a positive constant of proportionality, and *b* is a parameter in the range $b > 0$. It follows from this definition that a power law distribution is likely to be a good model for a set of empirical data if the plot of $\log(f(x))$ against $\log(x)$ resembles a straight line. It is quite amazing what a wide variety of variables in the physical, biological and social sciences has now been usefully modelled by a power law.

As everyone knows, the growth in the number of servers on the internet since the mid-1990s has been spectacular. In studies of the network of websites and of the behaviour of internet users at a point in time (say, a particular month), several statistics have turned up that have markedly L-shaped frequency distributions. Two examples are the number of links across the internet that point to a particular website, and the number of visits to a particular website. Everyday experience confirms that there are many websites which are very sparsely linked to, while a very few websites are pointed to from hundreds of thousands of other locations. Similarly, there are many websites which receive hardly any visits over a particular month, while a very few receive millions of hits.

In the former context, some successes and failures of a power law model are displayed in Pennock *et al.* (2002), online at [24.6]. In the latter context, a numerical example of fitting a power law model to an empirical frequency distribution can be found in Adamic (2002?), online at [24.7].

Question 24.4

George Zipf (1902–1950) was a linguistics professor at Harvard, whose initial use of statistics was for modelling word frequencies in English prose. In this connection, the statistical model that now bears his name proved very successful. Good insight into this aspect of his work is given by Alexander *et al.* (1998). Zipf went on to seek other variables whose frequency distribution he could represent well with the same probability model. One of these variables is the size of cities.

Clearly, the frequency distribution of a nation's cities by population size is likely to be highly skewed. The degree of skewness will probably differ in every country, so a model that would be suitable for all countries would probably require multiple shape parameters. Is there a way to view the modelling problem so that it needs very few parameters? Zipf found a way: his model has only one parameter. His approach was to transform any such skewed frequency distribution into a strictly L-shaped distribution. Then he could try to fit a model in the mathematical form of a power law. His solution was to use city rank as a *proxy* measure of frequency, rather than using the actual frequency (i.e. number of cities having a particular population size).

To follow Zipf's method, first rank the cities *in diminishing order* of population, then construct a 'frequency' distribution, the variable being population size and the proxy for frequency being city rank. For example, if r cities are ranked 1, 2, ... R, the smallest population value (on the X-axis) will correspond to the value R (on the Y-axis), the next smallest population value will correspond to $(R - 1)$, and so on. Do you see that using city ranks in this way produces exactly the same L-shaped distribution as constructing a *cumulative* frequency distribution of *number* of cities having a particular population size *or more*?

Zipf's Law states that the empirical distribution of city size, p, according to city rank, $r(p)$, is to be modelled by the power law $r(p) = ap^{-b}$, where a is a positive constant of proportionality and b is a parameter, with $b > 0$.

(We should mention that some writers present the Zipf model with the axes interchanged, so that city rank is on the X-axis and population is on the Y-axis. Adamic (2002?), online at [24.7], does this. She then explains how, in her formulation, the power law model and the Zipf Law model are algebraically related.)

In early empirical studies of Zipf's Law applied to cities, it was repeatedly found that b was very close to 1.0. This prompted some writers to claim that, in the city size context, the value $b = 1$ is some kind of global regularity! There has been much speculation in the statistical literature about why such a regularity should even be expected, let alone how it can be explained. Gabaix (1999) gives an informative perspective, comparing past explanations with a new one of his own. It is a technically advanced article, but its leading points are clearly made. There are now dozens of published studies of Zipf's Law in this application, though not all of them find that b is close to one. Soo (2005) offers an excellent overview.

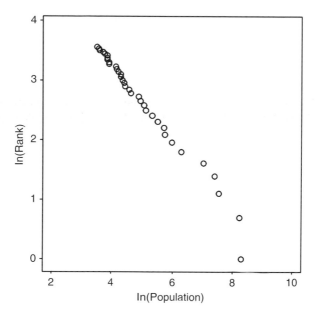

Figure 26.9 An illustration of Zipf's Law – Australia 2011.

Are you curious to know whether Zipf's Law is yet another confirmation of Stigler's Law of Eponymy (see Chapter 23)? Well, in its application to the size of cities it seems that it is. Zipf was anticipated in 1913 by Auerbach, whose work is cited, with some background, by Diego Rybski, online at [24.8].

Figure 26.9 shows a plot of *ln*(city rank) against *ln*(city population) for the 35 largest Australian cities, using the latest available census data from 2011. The scatter is clearly close to linear, as Zipf's Law predicts. Indeed, the value of the coefficient of determination for the fitted least squares regression line is 0.98. However, the estimated slope here is −0.63, and this value is statistically significantly different from −1.

Question 24.5

The two French researchers are Michel and Francoise Gauquelin (MG and FG). MG (1928–1991) and FG (1929–2007) each had an interest in astrology from an early age. Both before and after his marriage, MG pursued a lifelong agenda of statistical studies of correlations between certain configurations of the stars and planets in the heavens, and real-life events on Earth. Though critical of the often fatuous or bogus claims of astrologers, he proposed that if such a correlation were shown to have strong statistical significance in large samples of observations, it should be at least tentatively accepted by the orthodox scientific community as plausible evidence of a previously unrecognised true *causal* astrological relation.

MG first wrote about the 'Mars effect' in 1955 in his book *L'Influence des Astres* (The Influence of the Stars). The subsequent convoluted history of multiple attempts to sustain and to demolish the statistical evidence underpinning this supposedly real phenomenon has been traced over 40 years by Nienhuys (1997), online at [24.9]. An essay by Ertel and Irving (1999?), online at [24.10], dissents in minute detail from Nienhuys' commentary. Clearly, this has been an intensely contested proposition, though it never much engaged the scientific community at large. With MG's death in 1991, his claim is now hardly mentioned. There seems to be no basis here for a probability law.

References

Print

Alexander, L., Johnson, R. and Weiss, J. (1998). Exploring Zipf's Law. *Teaching Mathematics and its Applications* **17**, 155–158.

Fewster, R.M. (2009). A simple explanation of Benford's Law. *The American Statistician* **63**, 26–32.

Gabaix, X. (1999). Zipf's Law for cities: an explanation. *The Quarterly Journal of Economics* **114**, 739–767.

Nigrini, M. J. (1999). I've got your number. *Journal of Accountancy* **187**, 79–83.

Nigrini, M. J. (2012). *Benford's Law: Applications for Forensic Accounting, Auditing, and Fraud Detection*. Wiley.

Soo, K.T. (2005). Zipf's Law for cities: a cross country investigation. *Regional Science and Urban Economics* **35**, 239–263.

Online

[24.2] Poincaré, H. (1912). *Calcul des Probabilités*. Gauthier-Villars, 2nd edition, page 171. At: http://visualiseur.bnf.fr/ CadresFenetre?O=NUMM-29064&I=177&M=tdm

[24.3] Hill, T.P. (1995). A statistical derivation of the significant-digit law. *Statistical Science* **10**, 354–363. At https://projecteuclid.org/euclid. ss/1177009869

[24.4] Hill, T.P. (1998). The first digit phenomenon. *American Scientist* **86**, 358–363. At: http://people.math.gatech.edu/~hill/publications/ PAPER%20PDFS/TheFirstDigitPhenomenonAmericanScientist1996.pdf

[24.5]   Walthoe, J. (1999). Looking out for number one. *Plus Magazine*. At: http://plus.maths.org/issue9/features/benford/

[24.6]   Pennock, D.M. *et al.* (2002). Examples for: Winners don't take all: characterizing the competition for links on the web. At: http://modelingtheweb.com/example.html

[24.7]   Adamic, L.A. (2002?). Zipf, power-laws, and Pareto – a ranking tutorial. At: http://www.hpl.hp.com/research/idl/papers/ranking/ranking.html

[24.8]   Auerbach, F. (1913). Das Gesetz der Bevölkerungskonzentration. *Petermanns Geographische Mitteilungen* **59** (1), 73–76. Cited at: http://diego.rybski.de/files/RybskiD_EnvPlanA_2013.pdf

[24.9]   Nienhuys, J.W. (1997). The Mars Effect in retrospect. *Skeptical Inquirer* **21**(6), 24–29. At: http://www.skepsis.nl/mars.html.

[24.10] Ertel, S. and Irving, K. (1999?). The Mars Effect in prospect. At: http://www.planetos.info/dissent.html. There is more at http://www.planetos.info.

Answers – Chapter 25

Question 25.1

The mediaeval English Benedictine monk was the Venerable Bede (673–735), writing in the first chapter of his book *De temporum ratione* (*On the reckoning of time*), published in 725. His system of dactylonomy (Greek: *daktylos* finger + *nomos* law) or finger counting is able to represent numbers up to 9999. It was based on methods used by Arabs and Romans over many thousands of years. In his blog, *Laputan Logic*, John Hardy reproduces a 15th century illustration of Bede's finger counting system (see [25.4]). How Bede's system works in practice is explained on pages 201–207 of Karl Menninger's book *Number Words and Number Symbols – A Cultural History of Numbers* (1969). For more on this and other historical counting systems, see chapter 3 of Ifrah (1998).

Question 25.2

From 1879 onwards, Herman Hollerith (1860–1929) experimented – first at the US Census Bureau, and then at the Massachusetts Institute of Technology – with punched paper tape and, later, with punched cards, as ways of recording data that could be read and tabulated by machine. In 1887, he received a patent for an electric punched card reader. By 1890, he had devised an Electric Tabulating System, comprising punching, reading, sorting and tabulating machines. These machines were used to prepare the results of the 11th census of the United States in 1890. This work was completed in one-eighth of the time needed for obtaining manually the results of the previous census in 1880. Pictures of Hollerith's machines can be seen online at [25.5]. An interesting article on Hollerith's system by Mark Howells, with further pictures, is online at [25.6].

Question 25.3

A quincunx is any pattern of five objects arranged as in the five-spot on a die (with an object at each corner of a square and one in the middle). 'Quincunx' is the Latin word meaning 'five-twelfths' ('quinque' is 'five' and 'uncia' is 'one-twelfth'). The Roman quincunx was a coin worth five-twelfths of an 'as' (another coin). It had such a five-spot pattern on its reverse.

In statistics, the quincunx appears in a device to demonstrate the normal approximation to the binomial distribution. Such a device was first shown by Francis Galton in London in 1874. Several rows of pins were driven horizontally into a vertical backing board in the quincunx pattern. In use, small steel balls were allowed to cascade from an upper hopper through this array of pins to produce a symmetric histogram in several bins along the bottom of the board. A photo of Galton's device (he called the whole thing a 'quincunx') is included in the historical discussion in Stigler (1986), pages 275–281. A diagrammatic representation is given online at [25.7], and a video simulation (one of many on the web) is at [25.8].

Question 25.4

The Musical Dice Game (*Musikalisches Würfelspiel*) was published in Germany in 1793 and attributed by the publisher to Wolfgang Amadeus Mozart. Successive rolls of two dice select 16 individual bars of music from a compendium of bars. These selected bars, when assembled in the random order in which they were chosen, produce a harmonious minuet (a slow dance in waltz time, popular in the 17th and 18th centuries) every time. More details can be found in Ruttkay (1997). There are several computer versions of the game (for instance, one by Chuang, online at [25.9]), where you can try out the process yourself.

Question 25.5

A photograph of people (actually, female university students) arranged in groups by their height is the first of four photos in Brian Joiner's (1975) article 'Living histograms'. A second photo shows a bimodal 'living histogram' – Joiner coined this term – composed of both male and female students. A paper by Schilling, Watkins and Watkins (2002) reproduces Joiner's bimodal photo, and also provides a photo and citation from 1914, as evidence that Joiner was not the first to think of this informal creative way of illustrating a histogram. This earlier photo was published by Albert Blakeslee, a US plant geneticist.

More recently, Robert Jernigan (online at [25.10]) has drawn attention to an even earlier 'living histogram' photo. That photo comes from an article by Charles Davenport, published in 1901. Its educational appeal was subsequently recognised by Willard Brinton, who included it on page 165 of his 1914 textbook *Graphic Methods For Presenting Facts*. The complete book, including this particular page, can be viewed online at [25.11].

References

Print

Davenport, C.B. (1901). The statistical study of evolution. *The Popular Science Monthly*, September.

Ifrah, G. (1998). *The Universal History of Numbers*. Harvill Press.

Joiner, B.L. (1975). Living histograms. *International Statistical Review* **43**, 339–340 (note: the photos follow page 340 on unnumbered pages).

Menninger, K. (1969). *Number Words and Number Symbols – A Cultural History of Numbers*. MIT Press (reprinted by Dover in 1992).

Ruttkay, Z. (1997). Composing Mozart variations with dice. *Teaching Statistics* **19**, 18–19.

Schilling, M.F., Watkins A.E. and Watkins, W. (2002). Is human height bimodal? *The American Statistician* **56**, 223–229.

Stigler, S. (1986). *The History of Statistics: the Measurement of Uncertainty Before 1900*. Harvard University Press.

Online

[25.4] http://www.laputanlogic.com/articles/2004/05/11-0001.html

[25.5] http://www.officemuseum.com/data_processing_machines.htm

[25.6] http://www.mattivifamily.com/Sources/Census_Images/hollerith_system.pdf

[25.7] https://commons.wikimedia.org/wiki/File:Galton_Box.svg#/media/File:Galton_Box.svg

[25.8] https://www.youtube.com/watch?v=PM7z_03o_kk

[25.9] http://sunsite.univie.ac.at/Mozart/dice/

[25.10] http://statpics.blogspot.com/2008/05/earliest-living-histogram.html

[25.11] https://archive.org/details/graphicmethodsfo00brinrich and http://www.archive.org/stream/graphicmethodsfo00brinrich#page/164/mode/2up

Index

A Panorama of Statistics: Perspectives, Puzzles and Paradoxes in Statistics, First Edition.
Eric Sowey and Peter Petocz.
© 2017 John Wiley & Sons, Ltd. Published 2017 by John Wiley & Sons, Ltd.
Companion website: www.wiley.com/go/sowey/apanoramaofstatistics